FURTHER EXCAVATIONS AT MOHENJO-DARO

FURTHER EXCAVATIONS AT MOHENJO-DARO

Being an official account of Archaeological Excavations at Mohenjo-daro carried out by the Government of India between the years 1927 and 1931

By

E. J. H. MACKAY, M.A., D. Litt., F.S.A.,

Late Special Officer for Exploration, Archaeological Survey of India

With Chapters by A. S. Hemmy, B. A., M.Sc., and by B. S. Guha, Ph.D., and P. C. Basu, M.Sc., M. B.

Volume II : Plates I—CXLVI.

GYAN

Published by

Gyan Publishing House
5, Ansari Road
Daryaganj, New Delhi-110002
Phone: 011-47034999, 9811692060
E-mail: books@gyanbooks.com

Distribution Network
gyanbooks.com
India, USA, Canada, UK, Australia, France

ISBN : 978-81-212-2624-0 (Set)
ISBN : 978-81-212-2626-4 (HB)
First Published, 1937

2nd Impression 2021

Printed at: Gyan Press, Delhi.

Further Excavations At Mohenjo-Daro, Vol. II
Author: E. J. H. Mackay

FURTHER EXCAVATIONS AT MOHENJO-DARO

Being an official account of Archaeological Excavations at Mohenjo-daro carried out by the Government of India between the years 1927 and 1931

By

E. J. H. MACKAY, M.A., D.Litt., F.S.A.,

Late Special Officer for Exploration, Archaeological Survey of India.

With Chapters by A. S. Hemmy, B.A., M.Sc., and by B. S. Guha, Ph.D., and P. C. Basu, M.Sc., M.B.

Volume II : Plates I—CXLVI.

Published by the Manager of Publications, Delhi
Printed by the Manager, Government of India Press, New Delhi.
1937.

PLATES.

PLATE

XXXIV. DK Area, G Section, Southern Portion. Lower Levels—
(*a*) Short Lane, between Blocks 11 and 12. Inter. III level. From S.
(*b*) Long Lane. Stairway between Blocks 7 and 10 (11). Inter. II level. From N. W.
(*c*) Loop Lane, between Blocks 11 and 12A. From S.
(*d*) Loop Lane, between Blocks 11 and 12A. From N.

XXXV. DK Area, G Section, Southern Portion. Lower Levels—
(*a*) Block 1. Furnaces in house II, room 33. Inter. II level. From S. W.
(*b*) Block 1. Well in court 1 of house I. Inter. III level. From N. W.
(*c*) Block 1. Pottery cistern in house IV, room 80. Inter. II level. From W.
(*d*) Block 1. Coping of well in house I, room 19. Late III level. From S. E.
(*e*) Block 1. Stairway in house I, room 19. Inter. II level. From W.
(*f*) Block 1. Oven in court of house IV. Inter. II level. From N.
(*g*) Block 1. House IV, room 28. Inter. I and II doorways in northern wall. From S.

XXXVI. DK Area, G Section, Southern Portion. Lower Levels—
(*a*) Block 1. Well in house IV, room 38. Inter. I level. From E.
(*b*) Block 1. Rooms 63-65 of house VIII in foreground. From N. W.
(*c*) Block 2. Circular pavement in house II, room 22. Inter. II level. From S. E.
(*d*) Block 1. House IV, rooms 24-26. Inter. II level. From W.
(*e*) Block 2. Houses I and II in foreground. From S. W.

XXXVII. DK Area, G Section, Southern Portion. Lower Levels—
(*a*) Block 3. Upper walls with apertures, house V, room 21. Inter. I level. From S. W.
(*b*) Block 7. Runnel and jar in doorway of house IV, room 50. Inter. III level. From E.
(*c*) Block 7. Outside wall of house VII. Inter. II level. From N.
(*d*) Block 5. Well with Inter. I and Late III pavements. From W.
(*e*) Block 7. House V, court 68. Inter. III level. From N. W.
(*f*) General view of excavations. Blocks 2 and 3 in foreground. From S. E.
(*g*) Block 7. Well in house I, room 19. Inter. II level. From S.

XXXVIII. DK Area, G Section, Southern Portion. Lower Levels—
(*a*) Fore Lane. Repairs to house VI, Block 7. Inter. III level. From E.
(*b*) Long Lane. Chute outside Block 11, house III. Inter. III level. From S.
(*c*) Eastern end of Fore Lane. Inter. III level.
(*d*) Fore Lane. Runnel and jar outside Block 7. Inter. I level. From E.

XXXIX. DK Area, G Section, Southern Portion. Lower Levels—
(*a*) Block 9. Pavement and well in house VI. Inter. I level. From E.
(*b*) Block 9. Northern wall of house VII. Inter. II level. From S.
(*c*) Block 9. Pavements in house III. Inter. I level. From W.
(*d*) Block 9. Pipe drain in house VI, rooms 33-35. Inter. I level. From S. W.

PLATE

(e) Block 6A. Brick pier in rooms 38, 40. Late III period. From N.

(f) Block 9A. Cooking-place in house VIII, room 64. Late III period. From S. W.

XLVI. DK Area, G Section, Southern Portion. Upper Levels—

(a) North-eastern corner of Block 6A, with First Street in foreground.

(b) Block 9A with house VIII in foreground. From W.

XLVII. DK Area, G Section, Southern Portion. Upper Levels—

(a) First Street. Late III level. From N. E.

(b) First Street, with part of eastern wall of Block 3 in foreground. Late III level. From S. W.

(c) Blocks 5 and 6, showing lane between. Late III level. From E.

(d) Block 6. Eastern sides of houses I and II. Late III level.

XLVIII. DK Area, G Section, Southern Portion. Upper Levels—

(a) Block 7. Stairway in house II, room 93. Late III level. From S. W.

(b) Drain in lane between Blocks 9 (VI) and 9A (V). Late II level. From E.

(c) Block 12. Drain aperture in eastern wall of room 93. Late II level. From W.

(d) Late II drain between Blocks 12 and 12A. From S. W.

(e) Block 2. Cooking-places in court 21. Late Ib level. From N. E.

XLIX. DK Area, G Section, Southern Portion. Upper Levels—

(a) First Street, showing eastern side of Block 6. Late II level. From N. E.

(b) Block 9, house VI. Passage 31 in foreground. Late Ib level. From N. W.

(c) Block 9. Western side of house IV. Late Ib level. From N. W.

(d) Block 9. Eastern portion of house VI. From N. W.

(e) Block 1, house VII. Well in room 39. Upper pavement at Ia level. From N. E.

(f) Block 10, house II. Well and pavement in room 26. Late Ib level. From N.

L. DK Area, G Section, Southern Portion. Upper Levels—

(a) Block 9, house VI. Rooms 14-20 in background. Late Ia level. From S. W.

(b) Block 2. Pottery kiln in court 15. Late Ib level. From S. W.

(c) Block 9, house III. Late Ia drain. From S. E.

(d) Block 9. Pottery kilns in house II, court 21. Late Ia level. From E.

LI. DK Area, G Section, Southern Portion. Upper Levels—

(a) View of section before excavation. From S. W.

(b) Work commenced at Block 7, house I. From S. E.

(c) North-western corner of Block 6 in foreground, and First Street in distance.

(d) First Street. North-eastern corner of Block 3 in background.

PLATE

PLATE.

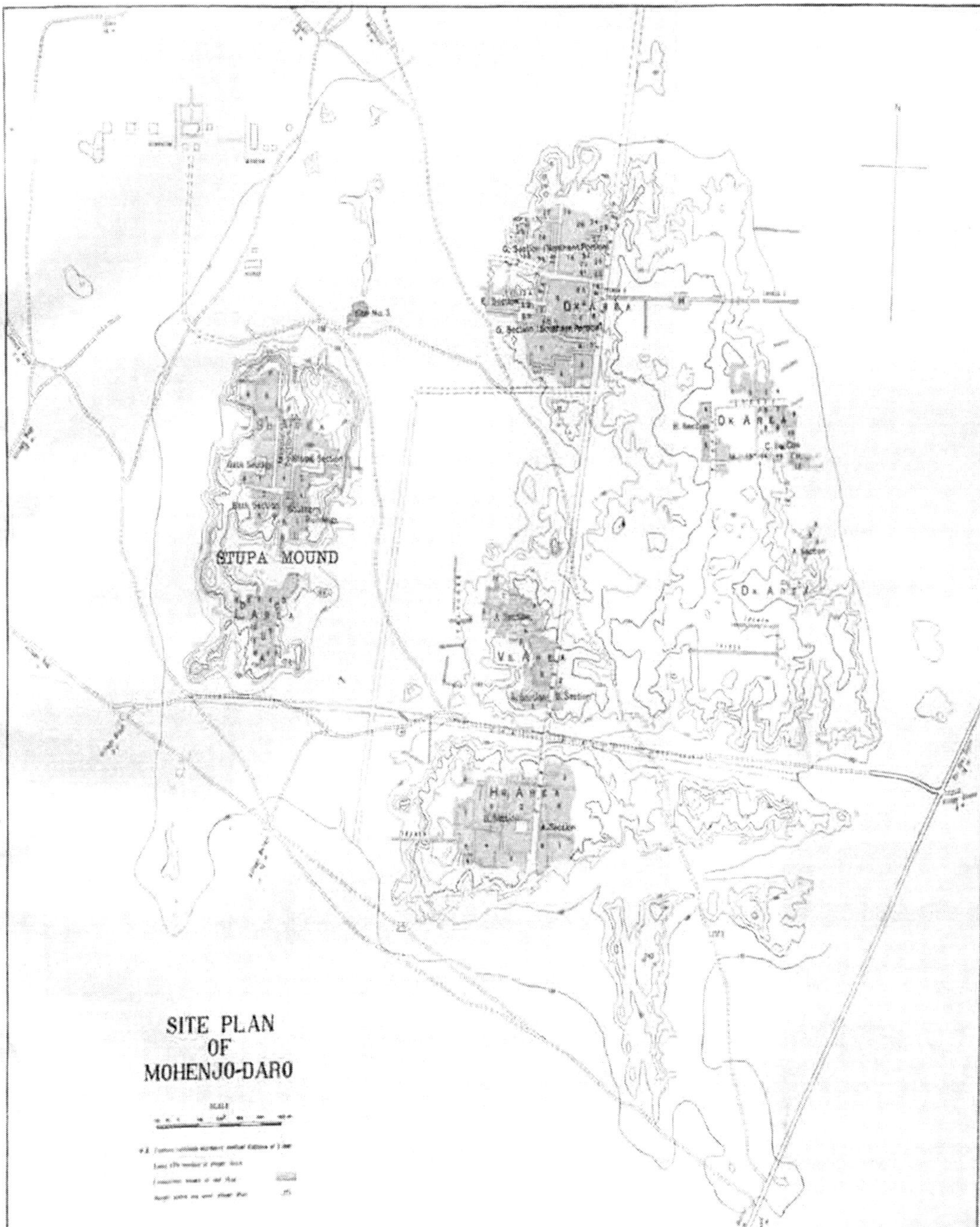
SITE PLAN
OF
MOHENJO-DARO
STUPA MOUND

MOHENJO-DARO, SD. AREA.

(a) Fenestrated court in Block 1. From South-East.

(b) Portion of Block 10 in foreground. From North-West.

(c) Block 1. South-eastern corner of fenestrated court.

(d) South-eastern corner of fenestrated courtyard showing late additions.

MOHENJO-DARO, SD. AREA.

(a) MAIN STREET FROM NORTH.

(b) ROOM 13, BLOCK 6. FROM WEST.

(c) MAIN STREET FROM SOUTH.

(d) ROOM 19, BLOCK 6. FROM EAST.

MOHENJO-DARO, SD AREA.

(a) LAYER OF GRAVEL BENEATH DRAIN IN ROOM 8, BLOCK 6. FROM WEST.

(b) DIVISION WALL AT NORTHERN END OF MAIN STREET. FROM SOUTH-EAST.

(c) PASSAGE BETWEEN BATHROOMS IN BLOCK 6. FROM EAST.

(d) BLOCK OF PAVING IN MAIN STREET. FROM SOUTH-WEST.

(e) CURVED DRAIN IN BLOCK 6 (S). FROM SOUTH.

(f) COLLAPSED DRAIN AT NORTHERN END OF MAIN STREET. FROM SOUTH.

(a) Blocks 1, 6 and 8: in Foreground. From N. W.

(b) Block 9. From S. E.

(c) Block 6. Passage and Group of Bathrooms. From W.

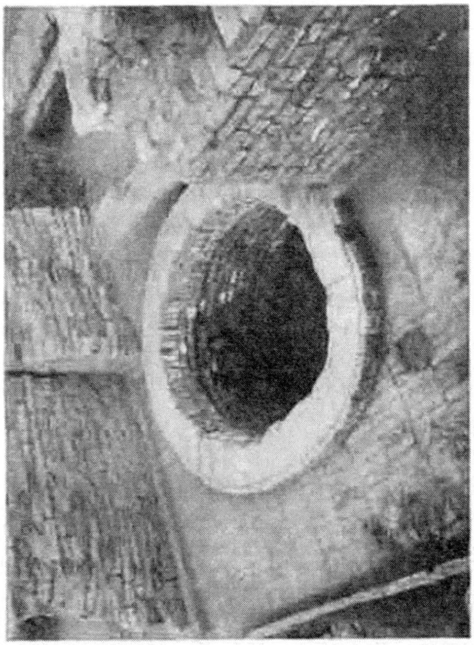

(d) Well in Chamber 18, Block 6. From S.

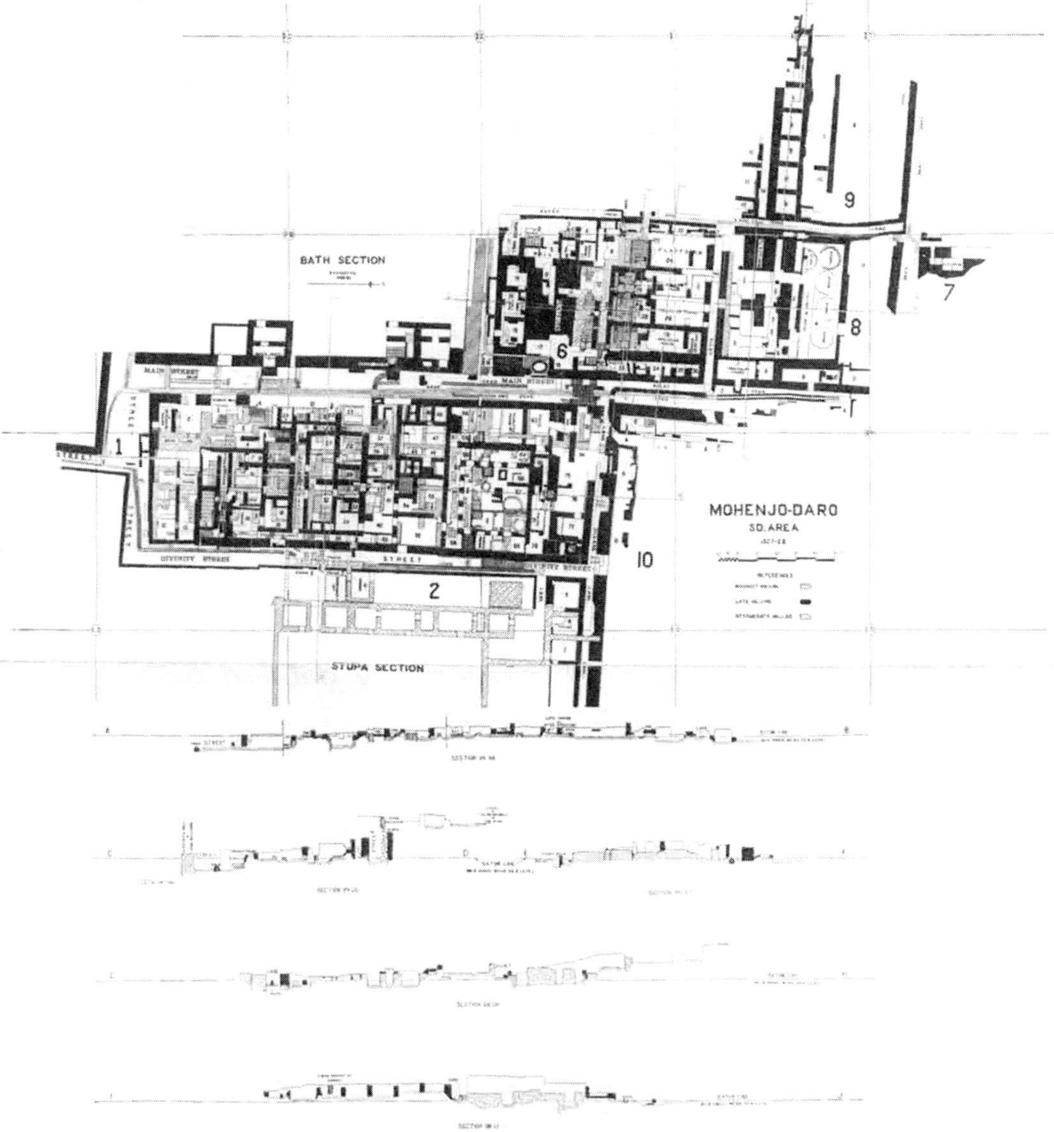
MOHENJO-DARO
SD. AREA
BATH SECTION
STUPA SECTION
MAIN STREET
STREET
1
2
6
7
8
9
10

Plate VI

MOHENJO-DARO

PLAN

of

LATE PERIOD BUILDING

in

SD AREA MIDDLE SECTION

BLOCK I

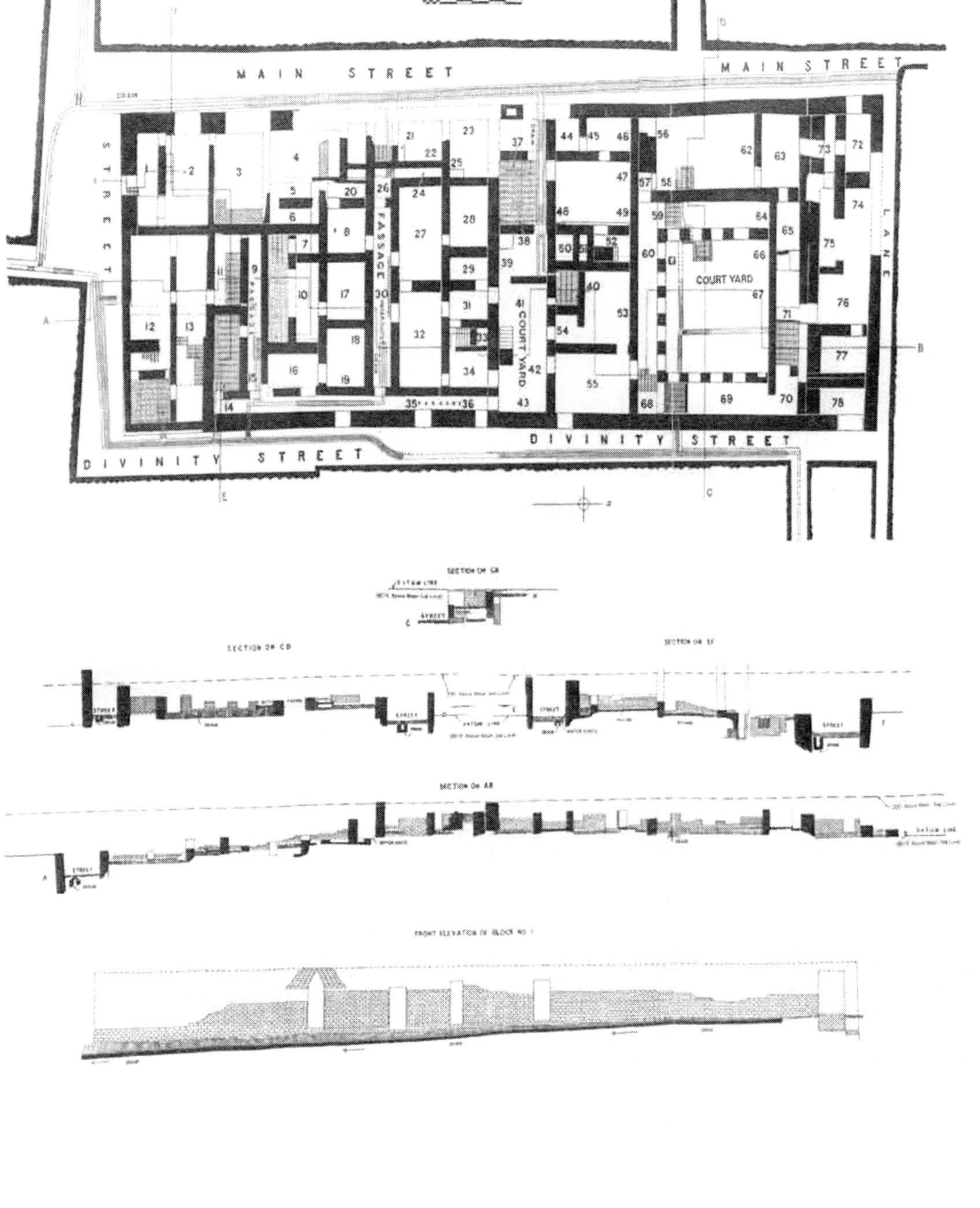

Plate VIII

SD. AREA, BATH SECTION, BLOCK 6.

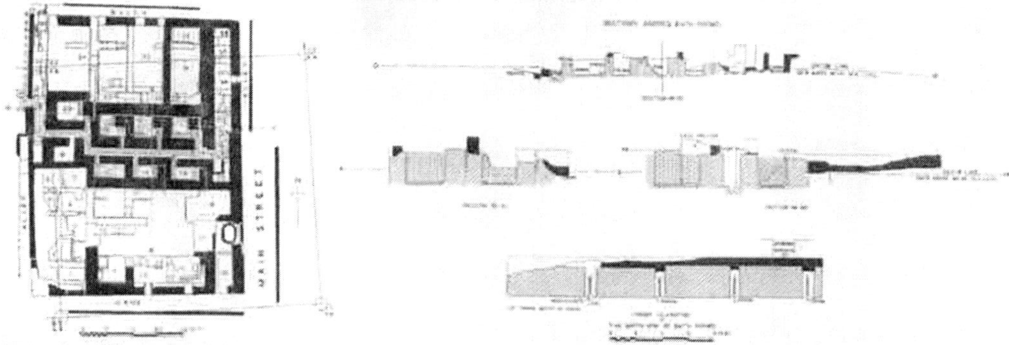

TRIAL TRENCHES (H); NORTH OF DK. AREA, G. SECTION.

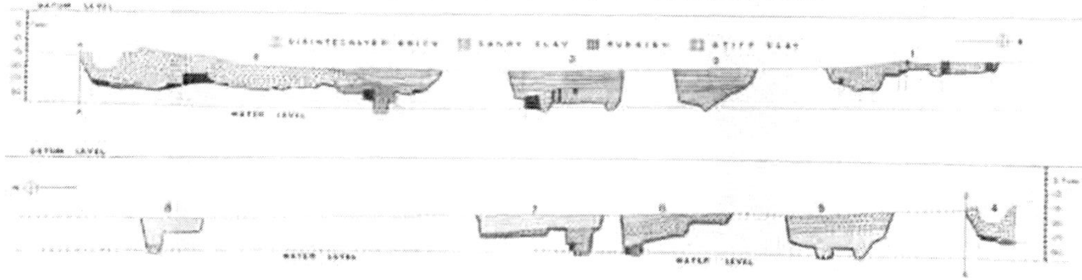

(a) GENERAL VIEW OF NORTHERN PORTION OF DK. AREA. PARTIALLY EXCAVATED FROM SOUTH-WEST.

(b) GENERAL VIEW OF EXCAVATIONS, NORTHERN PORTION, BLOCK 15, ROOMS 2 AND 3 IN FOREGROUND ON RIGHT.

(a) BLOCK 14, HOUSE III, ROOM 19. FROM N.W.

(b) CHUTE AND SOAK-PIT IN CENTRAL STREET. FROM N.

(c) CENTRAL STREET. LATE II LEVEL. FROM W.

(d) CENTRAL STREET. LATE II LEVEL. FROM E.

(e) GROUP A OF COPPER AND BRONZE TOOLS AND IMPLEMENTS.

(f) GROUP A OF COPPER AND BRONZE IN SITU.

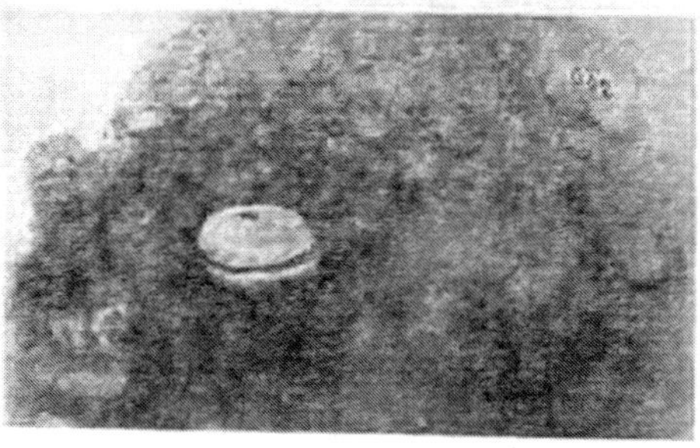

(g) GROUP B OF COPPER AND BRONZE IN SITU.

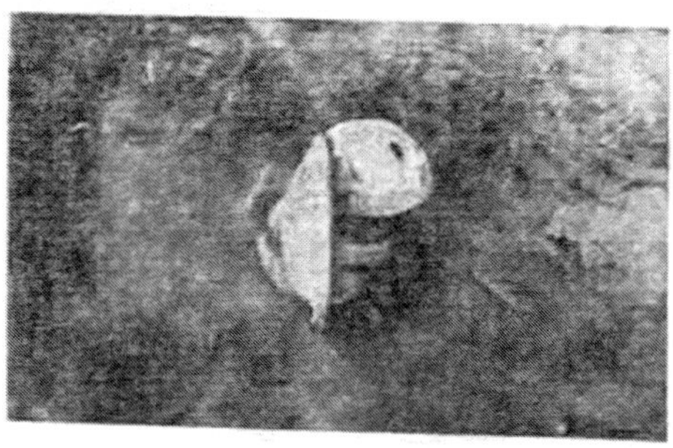

(h) GROUP C OF COPPER AND BRONZE.

MOHENJO-DARO, DK AREA, G. SECTION.

(a) West Street. Late Ia and b Levels. From South.

(b) West Street. Late II and III Levels. From South.

(c) West Street. Late Ia and b Levels. From North.

(d) West Street. Late II Level. From North.

(e) Block 18, Room 95. Pavement of Late Ib date. From North-West.

(f) Block 23, Building 1, Passage 8, 9. Pavement at Late II Level. From South-East.

MOHENJO-DARO, DK AREA, G. SECTION.

(a) Blocked door in West wall of Court 13, Block 23.

(b) Block 14, House III, Room 18. Bathroom or Privy. From West.

(c) Block 22, House I, Room 3. Pavement drain and pit of Late Ib Period. From South.

(d) Privy in House III, Block 23. From South.

(e) Block 15, House I, Room 3, Late II Level. From South-West.

(f) Block 18, Room 93. Pavement of Late Ia date From North-West.

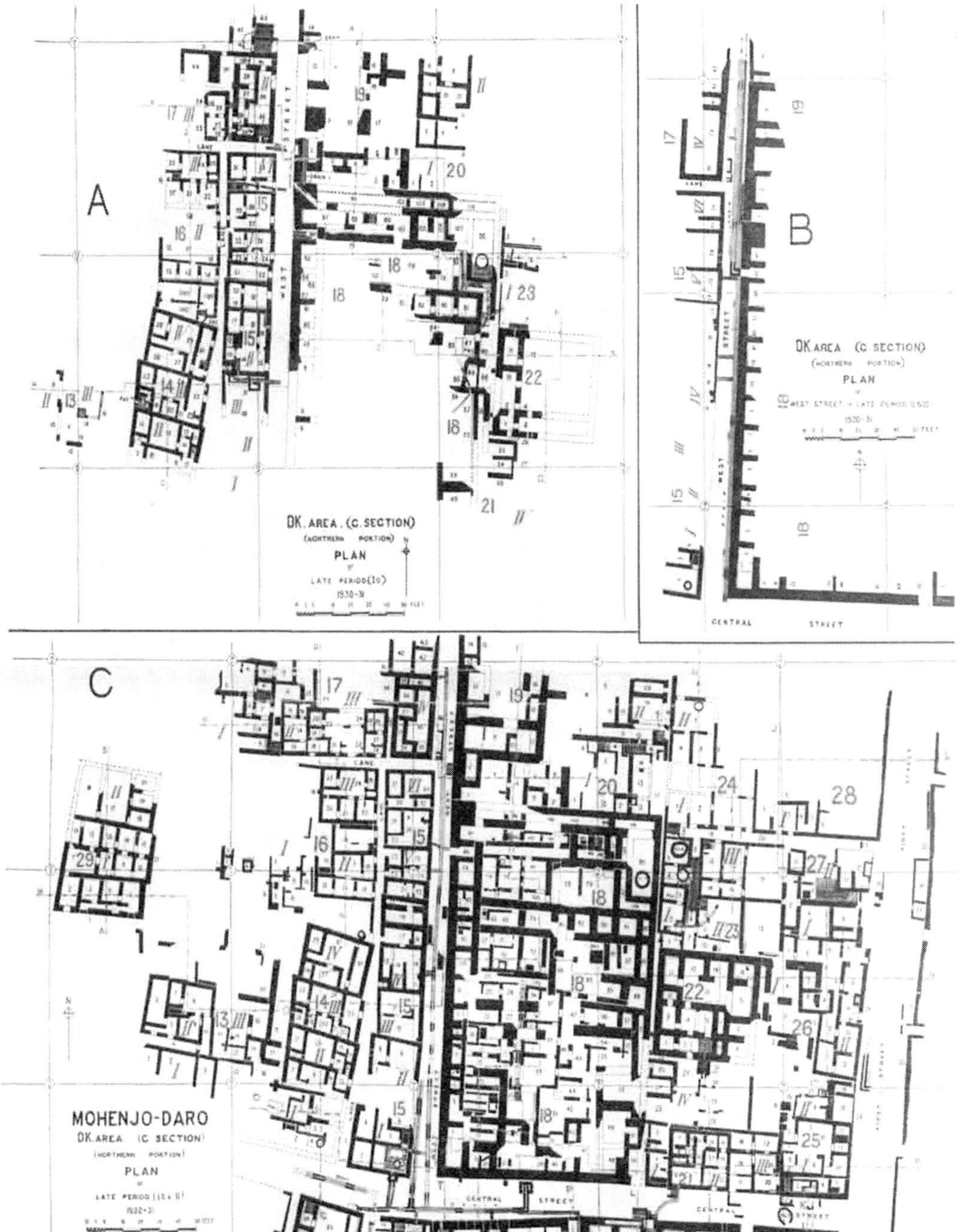
A
DK. AREA. (C. SECTION)
(NORTHERN PORTION)
PLAN
OF
LATE PERIOD (I.c)
1930-31
B
DK. AREA (C. SECTION)
(NORTHERN PORTION)
PLAN
OF
WEST STREET + LATE PERIOD
1930-31
CENTRAL STREET
WEST STREET
LANE
C
MOHENJO-DARO
DK. AREA (C. SECTION)
(NORTHERN PORTION)
PLAN
OF
LATE PERIOD (I.b + II)
1930-31
FIRST STREET
CENTRAL STREET

Plate XIII

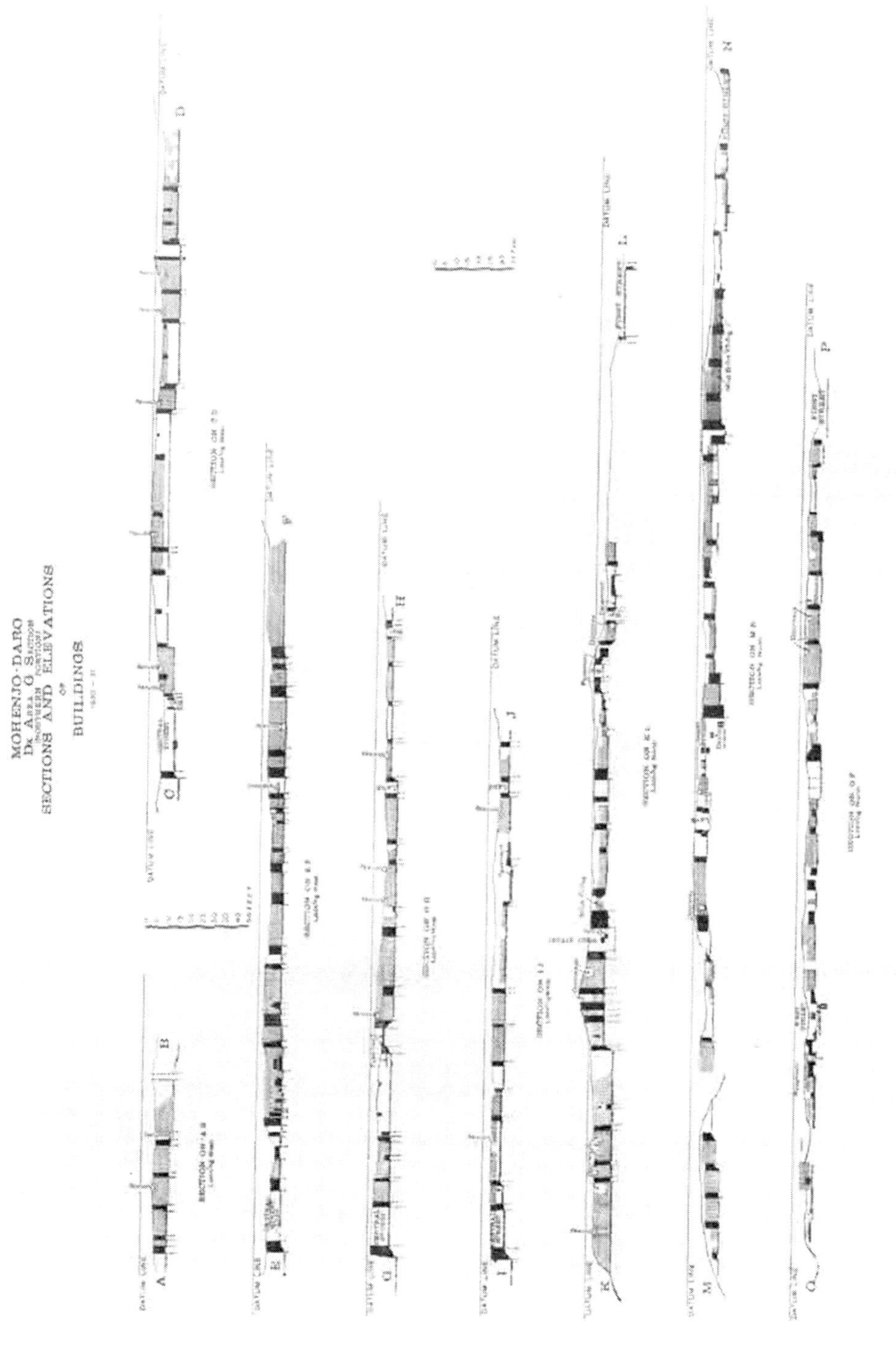
MOHENJO-DARO
DK AREA G SECTION
(NORTHERN PORTION)
SECTIONS AND ELEVATIONS
OF
BUILDINGS
A
B
C
D
E
F
G
H
I
J
K
L
M
N
O
P

(a) "Fore Lane." Early Drain and Buildings in Block 7. From West.

(b) "Fore Lane." Early Drain in Block 7. From West.

(c) Block 7. Early Structures excavated to 35 feet. From North-East.

(d) Block 7. Small Building at very Early Level. From South-West.

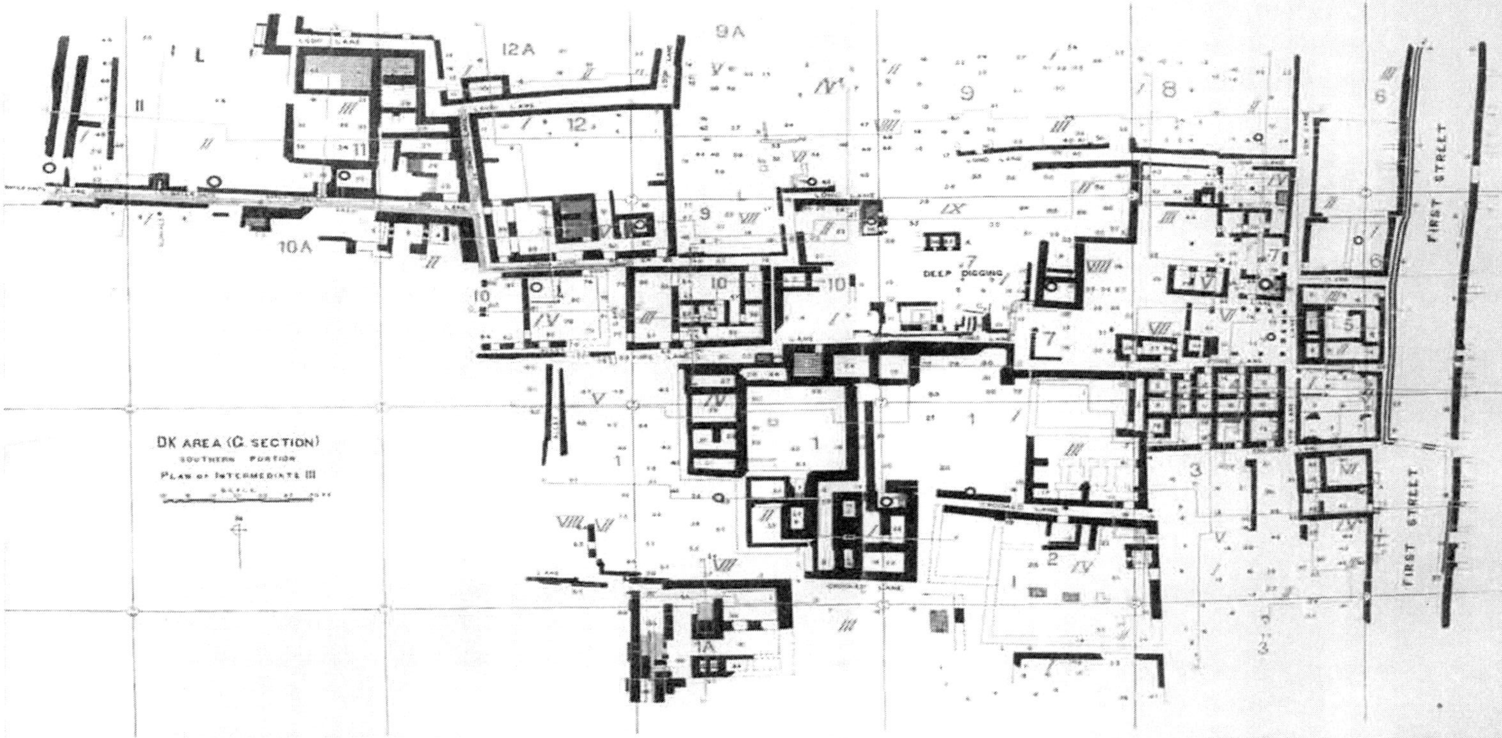
DK AREA (G SECTION)
SOUTHERN PORTION
PLAN OF INTERMEDIATE III
FIRST STREET
FIRST STREET
DEEP DIGGING
CROOKED LANE
12A
9A
10A
1A

PLATE XVII

MOHENJO-DARO
DK AREA
C. SECTION
PLAN
INTERMEDIATE II PERIOD
1928-29-30
FIRST STREET
FIRST STREET

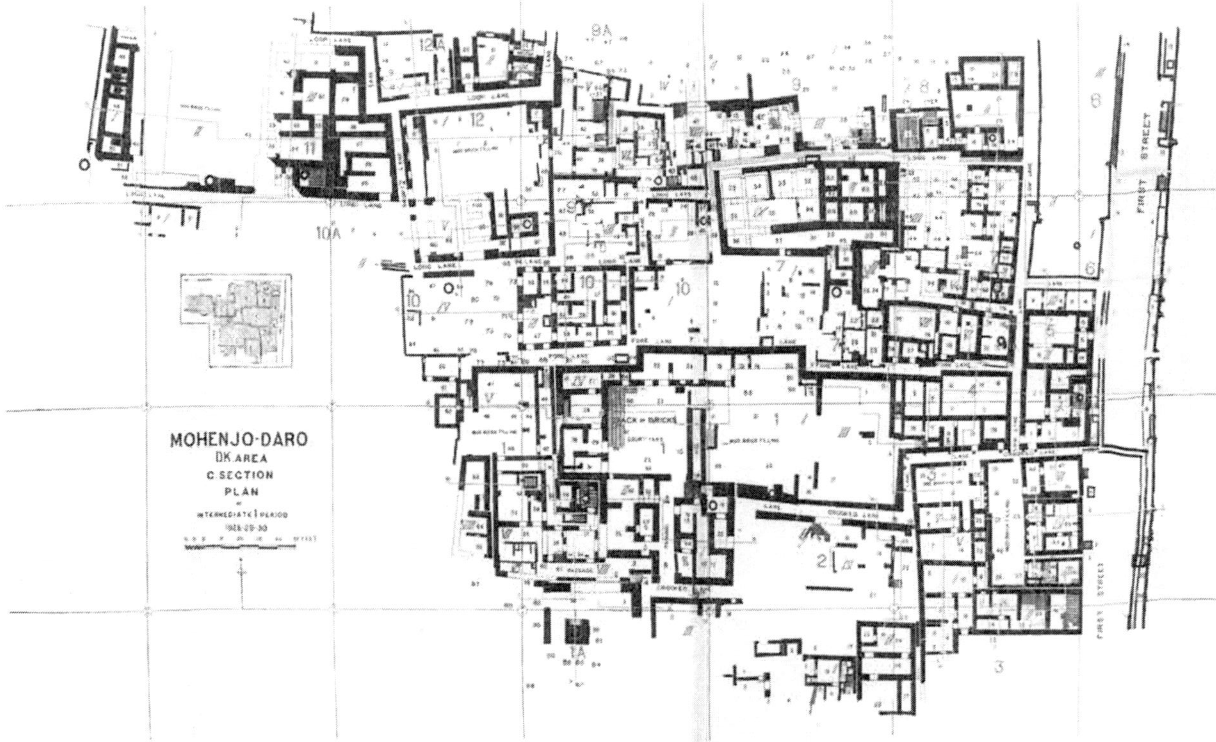
MOHENJO-DARO
DK AREA
G SECTION
PLAN
OF
INTERMEDIATE I PERIOD
1928-29-30
FIRST STREET
LOOP LANE
LONG LANE
FORE LANE
CROOKED LANE
STACK OF BRICKS
MUD BRICK FILLING
PASSAGE

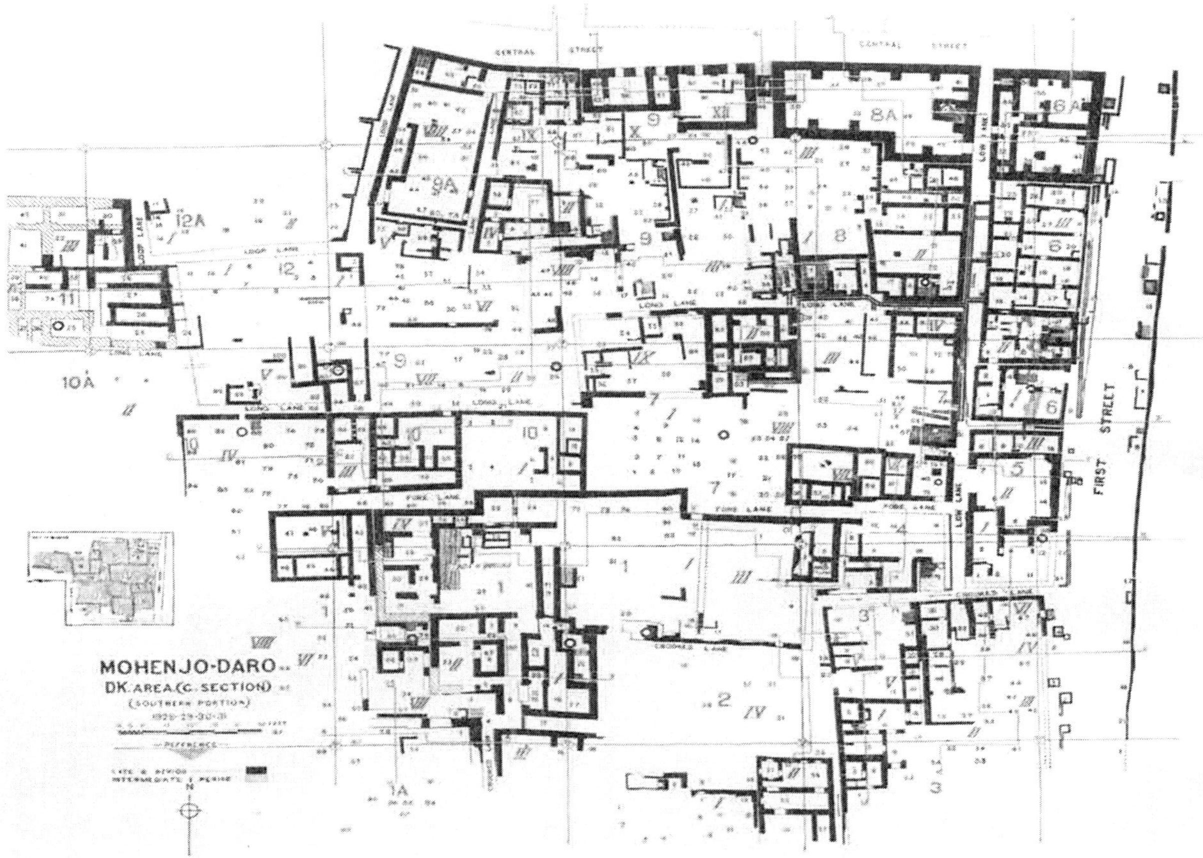
CENTRAL STREET
FIRST STREET
LOW LANE
LOOP LANE
LONG LANE
FORE LANE
CROOKED LANE
MOHENJO-DARO
DK. AREA (G. SECTION)
(SOUTHERN PORTION)
1928-29-30-31
REFERENCE
LATE II PERIOD
INTERMEDIATE I PERIOD
N

CENTRAL STREET
CENTRAL STREET
FIRST STREET
FIRST STREET
STREET ?
LONG LANE
LOOP LANE
CROOKED LANE
FORE LANE
MOHENJO-DARO
DK AREA — G SECTION
(SOUTHERN PORTION)
PLAN OF LATE I & II PERIODS
SCALE
N

MOHENJO-DARO DK AREA (G SECTION) SOUTHERN PORTION

SECTIONS AND ELEVATIONS OF BUILDINGS.

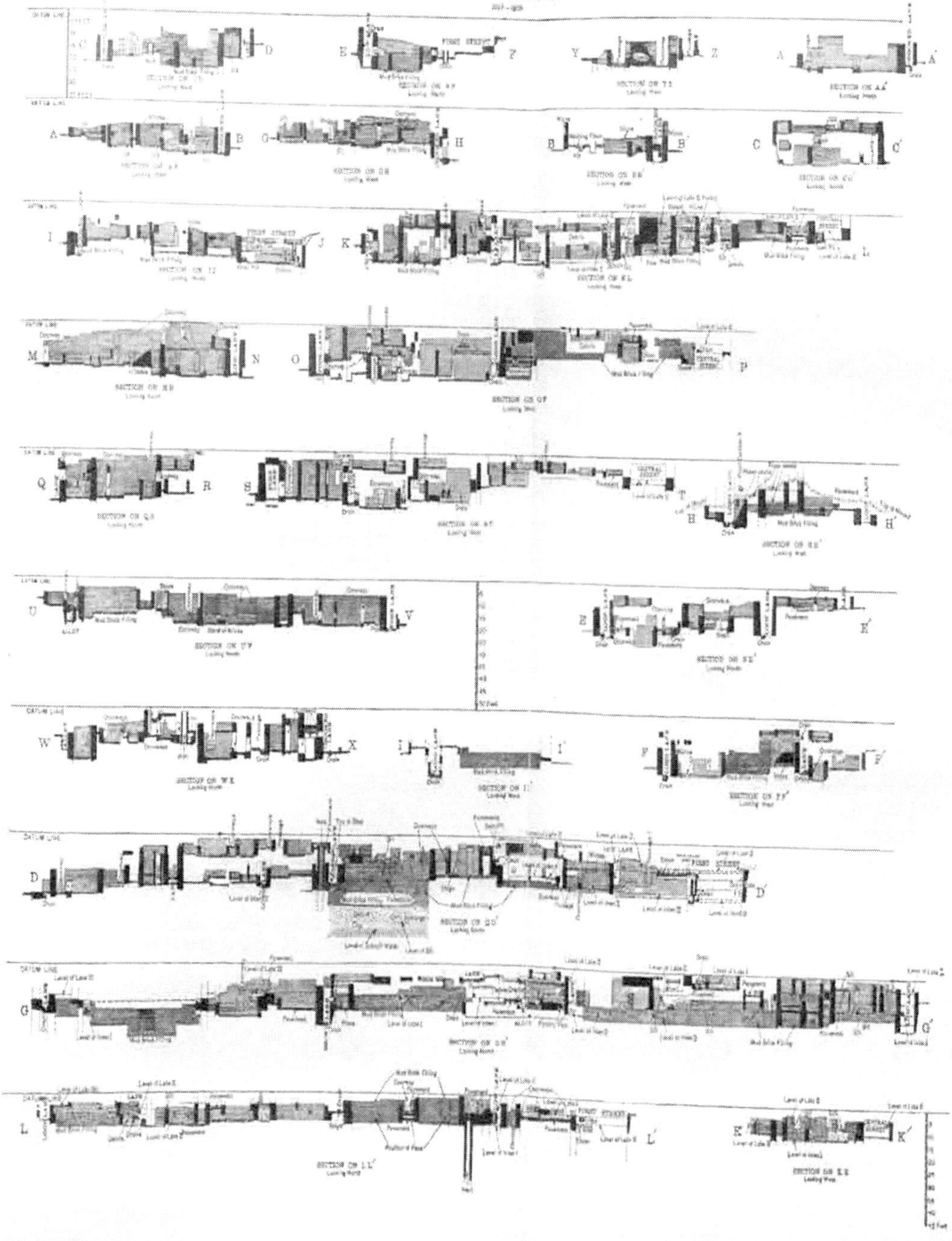

ELEVATIONS AND PLANS OF DETAILS.

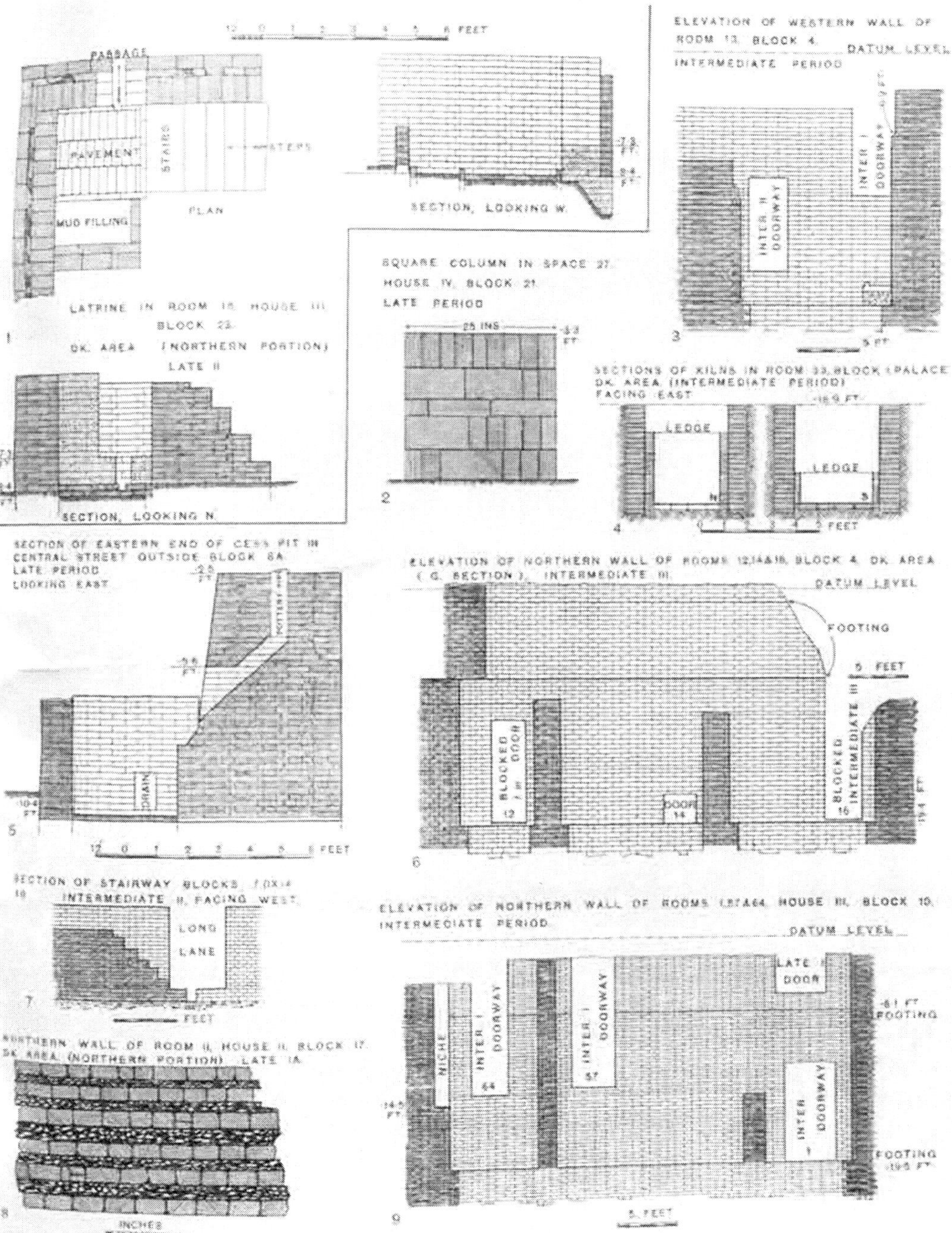

KILN, DRAINS, SEDIMENT PITS, ETC.

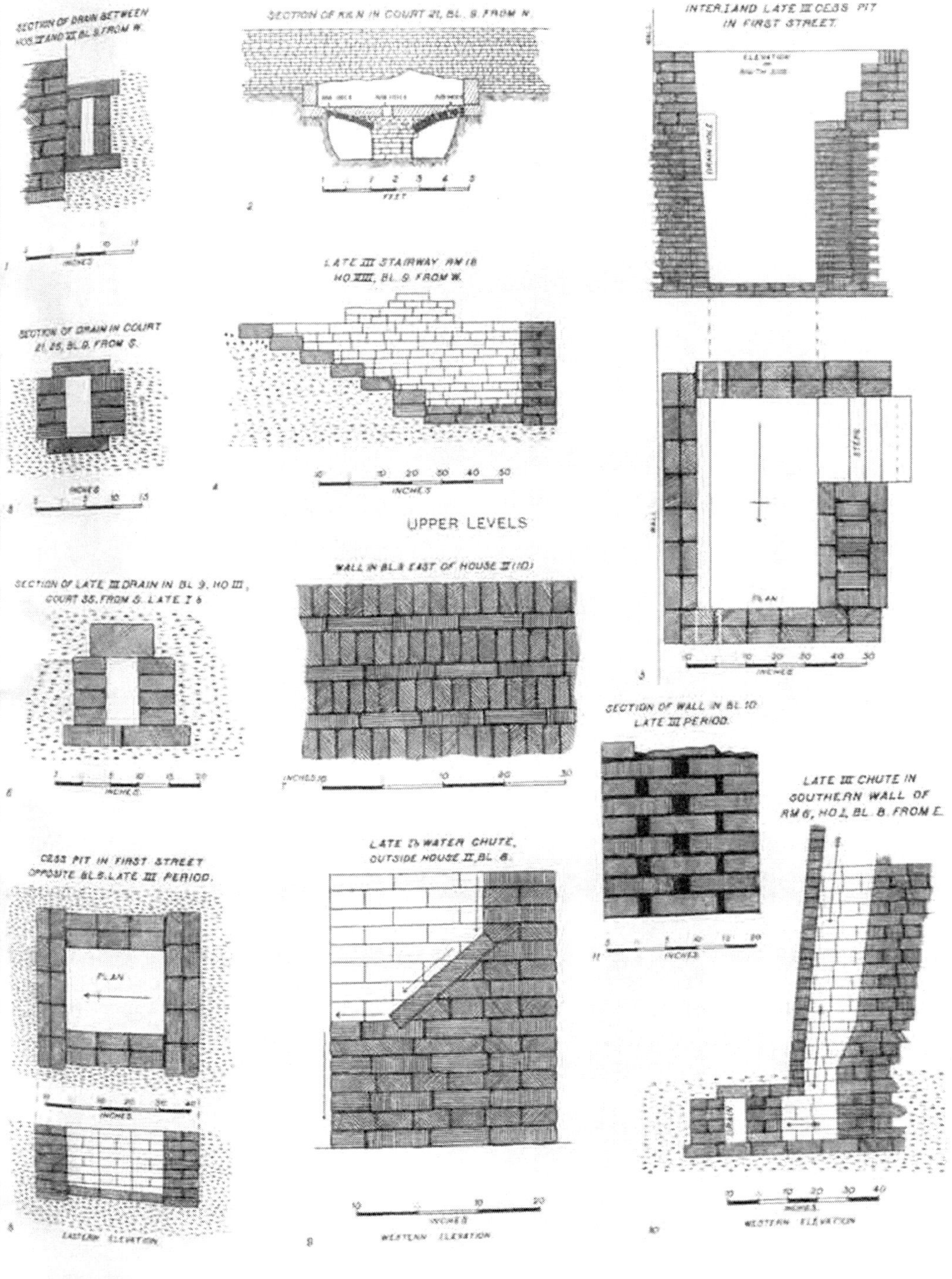

(a) "FIRST STREET"; TO SUGGEST ITS ANCIENT POPULATION. FROM SOUTH.

(b) WELL IN COURT IV, BLOCK 10 FROM SOUTH-WEST.

MOHENJO-DARO, DK. AREA, G. SECTION.

(a) FIRST STREET. INTER. I LEVEL. FROM NORTH.

(b) FIRST STREET. INTER. I LEVEL. FROM SOUTH.

(c) FIRST STREET. INTER. II LEVEL. FROM NORTH.

(d) FIRST STREET. INTER. II LEVEL. FROM SOUTH.

(e) FIRST STREET. INTER. III LEVEL. FROM NORTH.

(f) FIRST STREET. INTER. III LEVEL. FROM SOUTH.

(a). General View of Excavations showing Block 1A in foreground. Intermediate III Level. Facing North.

(b) Blocks 9 and 10, the former partially excavated. Facing South-East.

(a) Block 11 with Block 10A and Long Lane in Foreground. Intermediate III Level. Facing North-West.

(b) Block 12 with Passage 99 on left. Partially excavated. Intermediate III Level. Facing North.

(a) EASTERN COURT OF BLOCK 1, WITH EAST WING AND BLOCK 4 BEYOND. INTERMEDIATE II LEVEL. FACING NORTH-EAST.

(b) EASTERN COURT OF BLOCK 1. INTERMEDIATE II LEVEL. FACING SOUTH-EAST.

(a) Pot drain in First Street. Inter. III Level. From West.

(b) Block 7, House IV, Room 79. Inter. III Level. From West.

(c) First Street. Drain with trimmed bricks. Inter. I Level. From West.

(d) Chute and Drain in First Street. Inter. II Level. From East.

MOHENJO-DARO, DK. AREA, G. SECTION.

(a) BLOCK 4, ROOMS 14–16. INTER. III LEVEL. FROM SOUTH-EAST.

(b) BLOCK 4, ROOMS 14–16. INTER. III LEVEL. FROM SOUTH-WEST.

(c) "CROOKED LANE" SOUTH OF BLOCKS 4 AND 5. INTER. III LEVEL. FROM WEST.

(d) LOW LANE, EAST OF BLOCKS 4 AND 7. INTER. III LEVEL. FROM NORTH.

MOHENJO-DARO, DK. AREA, G. SECTION.

(a) BLOCK 12A HOUSE II, ROOMS 22-23 IN FOREGROUND. FROM NORTH-EAST.

(b) BLOCK 11, HOUSE III, ROOM 25 AT LEFT. FROM SOUTH.

(c) "LONG LANE" BETWEEN BLOCKS 9 AND 10 (II). INTER. II LEVEL. FROM WEST.

(d) "LONG LANE" BETWEEN BLOCKS 9 (VII) AND 10. FROM EAST.

(e) "SHORT LANE" BETWEEN BLOCKS 11 AND 12. FROM SOUTH.

MOHENJO-DARO. DK. AREA. G. SECTION.

(a) SKELETON NO. 6 IN BLOCK 10A. FROM SOUTH.

(b) SKELETON NO. 7 IN BLOCK 10A. FROM SOUTH.

(c) "LONG LANE" BETWEEN BLOCKS 10A AND 11. FROM EAST.

(d) "LONG LANE" BETWEEN BLOCKS 10A AND 11. FROM WEST.

(a) Entrance door of House III, Block 10. Inter. III Level, Facing E.

(b) Double Stairway in Block 1A. Inter. III Level. From North.

(c) "Fore Lane." South of House III, Block 10. Inter. III Level. From East.

(d) "Fore Lane." South of House III, Block 10. Inter. III Level. From West.

MOHENJO-DARO, DK. AREA, G. SECTION.

(a) "Short Lane." Between Blocks 11 and 12. Inter. III Level. From South.

(b) "Long Lane." Stairway between Blocks 7 and 10 (II). Inter. II Level. From North-West.

(c) "Loop Lane." Between Blocks 11 and 12A. Inter. II Level. From South.

(d) "Loop Lane." Between Blocks 11 and 12A. Inter. II Level. From North.

MOHENJO-DARO. DK. AREA. G SECTION.

(a) Kiln in Block 1, House II, Room 33. Inter. II Level. From South-West.

(b) Well in Block 1, Court I. Inter. III Level. From North-West.

(c) Bowl in Block 1, House IV, Room 80. Inter. II Level. From West.

(d) Well in Block H 1, House I, Room 19. Late III Level. From South-East.

(e) Stairway in Block 1, House I, Room 19. Inter. II Level. From West.

(f) Oven in Court of House IV, Block 1. Inter. II Level. From North.

(g) Block 1, House IV, Room 28. Inter. I and II Doorways. From South.

MOHENJO-DARO, DK AREA, G SECTION.

(a) Block 1, House IV, Room 38. Inter. I Level. From East.

(b) Block 1, Rooms 63-65 of House VIII in foreground. From North-West.

(c) Washing-floor in Block 2, House II, Room 22. From South-East.

(d) Block 1, House IV, Rooms 24-26. Inter. II Level. From West.

(e) Houses I and II of Block 2 in foreground. Inter. II Level. From South-West.

MOHENJO DARO. DK. AREA, G. SECTION.

(a) Apertures in walls of Block 3, House V, Room 21 Inter. I Period. From South-West.

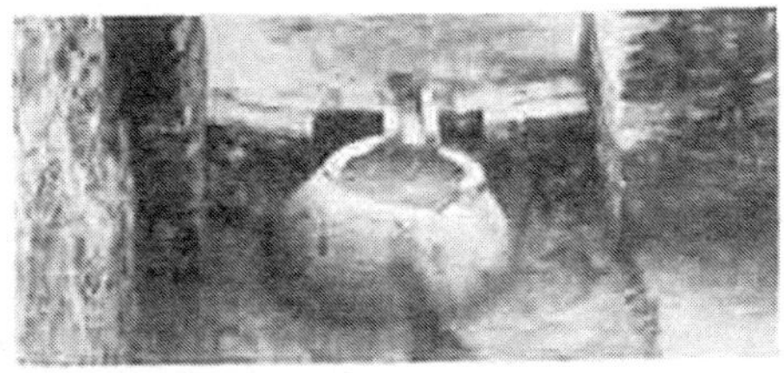

(b) Runnel and Jar in Block 7, House IV, Room 50. Inter. III Level. From East.

(c) Outside wall of Block 7, House VII, Inter. II Level. From North.

(d) Block 5, Room 12, Inter. I and Late III Levels. From West.

(e) Block 7, House V, Inter. III Level. From North-West.

(f) Blocks 2 and 3 in foreground, Facing North-West.

(g) Block 7, House I, Room 19. Inter. II Level. From South.

(a) Foundations of south side of House VI, Block 7. Inter. III Level. From East.

(b) Chute north side of lane between Blocks 10A and 11 (III).

(c) "Fore Lane." Foundations of south side of House VI, Block 7. From West.

(d) "Fore Lane." Chute outside Block 7. Inter. I Level. From East.

MOHENJO-DARO, DK. AREA, G. SECTION.

(a) BLOCK 9, HOUSE VI, ROOM 46, INTER. I LEVEL. FROM EAST.

(b). BLOCK 9, HOUSE VII, NORTH WALL. INTER. II PERIOD. FROM SOUTH.

(c) BLOCK 9, HOUSE III, ROOM 35. INTER. I LEVEL. FROM WEST.

(d) BLOCK 9, HOUSE VI, ROOM 34. INTER. I LEVEL. FROM SOUTH-WEST.

(e) BLOCKS 7, 9 AND 10, HOUSE VI OF BLOCK 9 IN RIGHT FOREGROUND. FROM NORTH-WEST.

(f) "LONG LANE" BETWEEN BLOCKS 7 AND 9 (III) INTER. II LEVEL. FROM WEST.

(a) FACADE OF HOUSE III, BLOCK 10. INTER. III LEVEL. FROM WEST.

(b) NORTH-EAST CORNER OF HOUSE III, BLOCK 10. INTER. III LEVEL.

(c) SOUTH WALL OF ENCLOSURE 83, HOUSE IV, BLOCK 10. INTER. III LEVEL.

(d) STAIRWAY IN BLOCK 1A. INTER. III LEVEL. FROM NORTH EAST.

MOHENJO-DARO, DK. AREA, G. SECTION.

(a) "Loop Lane" between Blocks 9A and 12A. From North.

(b) Well in Block 10, House II, Room 27. Inter. III Level. From West.

(c) Chute outside House 4, Block 10. Inter. II Level. From South-West.

(d) Well in Block 10, House II. Inter. III Level. From South-East.

(e) Block 9, House VII, "Long Lane" on left. Inter. I Level. From East.

(f) Block 10, House II, Room 27. Inter. II Level. From South-West.

(g) "Long Lane" and Block 10A in foreground. From South East.

(a) "FORK LANE" AND NORTH FACE OF GREAT WALL. INTER. II LEVEL. FROM N.W.

(b) BLOCK 12, ROOMS 90 AND 100. INTER. III LEVEL. FROM S.

(c) "LOOP LANE," BETWEEN BLOCKS 12 AND 12A. INTER. II LEVEL. FROM W.

(d) BLOCK 12, ROOMS 89, 90, 99 AND 100. INTER. III LEVEL. FROM N.E.

MOHENJO-DARO, DK. AREA, G. SECTION.

...ow Lane. Late II Cess-pits. From North.

(b) Block 8A, Room 42. Well and Pavement of Late II period. From East.

(c) Block 8A, Room 42. Skeletons on stairway of Late Ib date. From West.

(d) Block 8, House II, Room 50. Drain covered with large bricks. Late I date. From West.

(e) Block 9, House XII, Room 45. Washing Floor. Late Ib Level. From North.

(f) Block 9, House XII, Room 44. Well and Pavement of Late Ib date. From North East.

MOHENJO-DARO, DK. AREA, G. SECTION.

(a) [illegible] outside Room 17, House II, Block 8. Late Ib Level. From South

(b) Well in Chamber 19, House II, Block 8. Late III and II Levels. From South-East.

(c) "First Street." Late III Drainage System. From North.

(d) "Low Lane" showing Late III Drains. From North.

(e) Blocks 7 and 9: House IX of Block 7 in middle distance. Late III Level. From North.

(f) Block 9, House VI. Late III Structure. From North-West.

MOHENJO-DARO, DK AREA, G. SECTION.

(a) FIRST STREET. DRAIN AND CESS PITS OF INTERMEDIATE I DATE. FROM NORTH-WEST.

(b) BLOCK 8A, ROOMS 33, 34. FROM SOUTH-WEST.

(c) BLOCK 8A, ROOMS 37, 40. LATE II LEVEL. FROM SOUTH-WEST.

(d) BLOCK 8A, ROOMS 33,34. LATE II LEVEL. FROM SOUTH-EAST.

(e) PIER IN BLOCK 6A, ROOMS 38, 40. LATE III PERIOD. FROM NORTH.

(f) CHANNELS BENEATH PAVING OF ROOM 64, HOUSE VIII, BLOCK 9A. LATE III LEVEL. FROM SOUTH-WEST.

(a) NORTH-EASTERN CORNER OF BLOCK 6A WITH FIRST STREET IN FOREGROUND.

(b) BLOCK 9A WITH HOUSE VIII IN FOREGROUND. FROM WEST.

MOHENJO-DARO, DK. AREA, G. SECTION.

(a) "FIRST STREET." LATE III LEVEL. FROM NORTH EAST.

(b) "FIRST STREET." PART OF EASTERN WALL OF BLOCK 3 IN FOREGROUND. LATE III LEVEL. FROM SOUTH-WEST.

(c) LANE BETWEEN BLOCKS 5 AND 6. LATE III LEVEL. FROM EAST.

(d) BLOCK 6. EASTERN SIDES OF HOUSES I AND II LATE III LEVEL.

(a) BLOCK 7, HOUSE II. LATE III STAIRWAY IN ROOM 93. FROM SOUTH-WEST.

(b) DRAIN BETWEEN BLOCKS 9 (VI) AND 9A. LATE II LEVEL. FROM EAST.

(c) BLOCK 12. DRAIN APERTURE IN EAST WALL OF ROOM 93. LATE II LEVEL. FROM WEST.

(d) LATE II DRAIN BETWEEN BLOCKS 12 AND 12A. FROM SOUTH-WEST.

(e) COOKING PLACES. BLOCK 2, COURT 21. LATE Ib LEVEL. FROM NORTH-EAST.

MOHENJO-DARO, DK AREA G SECTION.

(a) FIRST STREET SHOWING EAST SIDE OF BLOCK 6, LATE II LEVEL. FROM NORTH-EAST.

(b) BLOCK 9, HO. VI, PASSAGE 31 IN FOREGROUND. LATE IB LEVEL. FROM NORTH-WEST.

(c) BLOCK 9, WESTERN SIDE OF HOUSE IV. LATE IB LEVEL. FROM NORTH-WEST.

(d) BLOCK 9, EASTERN PORTION OF HOUSE VI. LATE IB LEVEL. FROM NORTH-WEST.

(e) BLOCK 1, HOUSE VII. WELL IN ROOM 39. TOP PAVEMENT, LATE IA LEVEL. FROM NORTH-EAST.

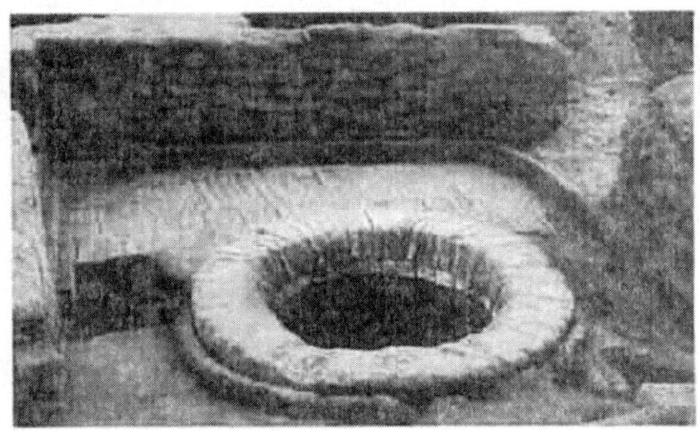

(f) BLOCK 10, HOUSE II. WELL AND PAVING OF LATE IB LEVEL. FROM NORTH.

(a) Block 9, House VI, Rooms 14-20 in background. Late Ia Level. From South-West.

(b) Kiln in Block 2, Court 15. Late Ib Level. From South-West.

(c) Block 9, House III. Late Ia Drain. From South-East.

(d) Two kilns in Block 9, House II, Court 21. Late Ia Level. From East.

(a) DK AREA, G. SECTION, SOUTHERN PORTION BEFORE EXCAVATION. FROM SOUTH-WEST.

(b) EXCAVATION OF HOUSE I, BLOCK 7. FROM SOUTH-EAST.

(c) NORTH-WESTERN CORNER OF BLOCK 6 IN FOREGROUND.

(d) FIRST STREET. NORTH-EASTERN CORNER OF BLOCK 3 IN BACKGROUND.

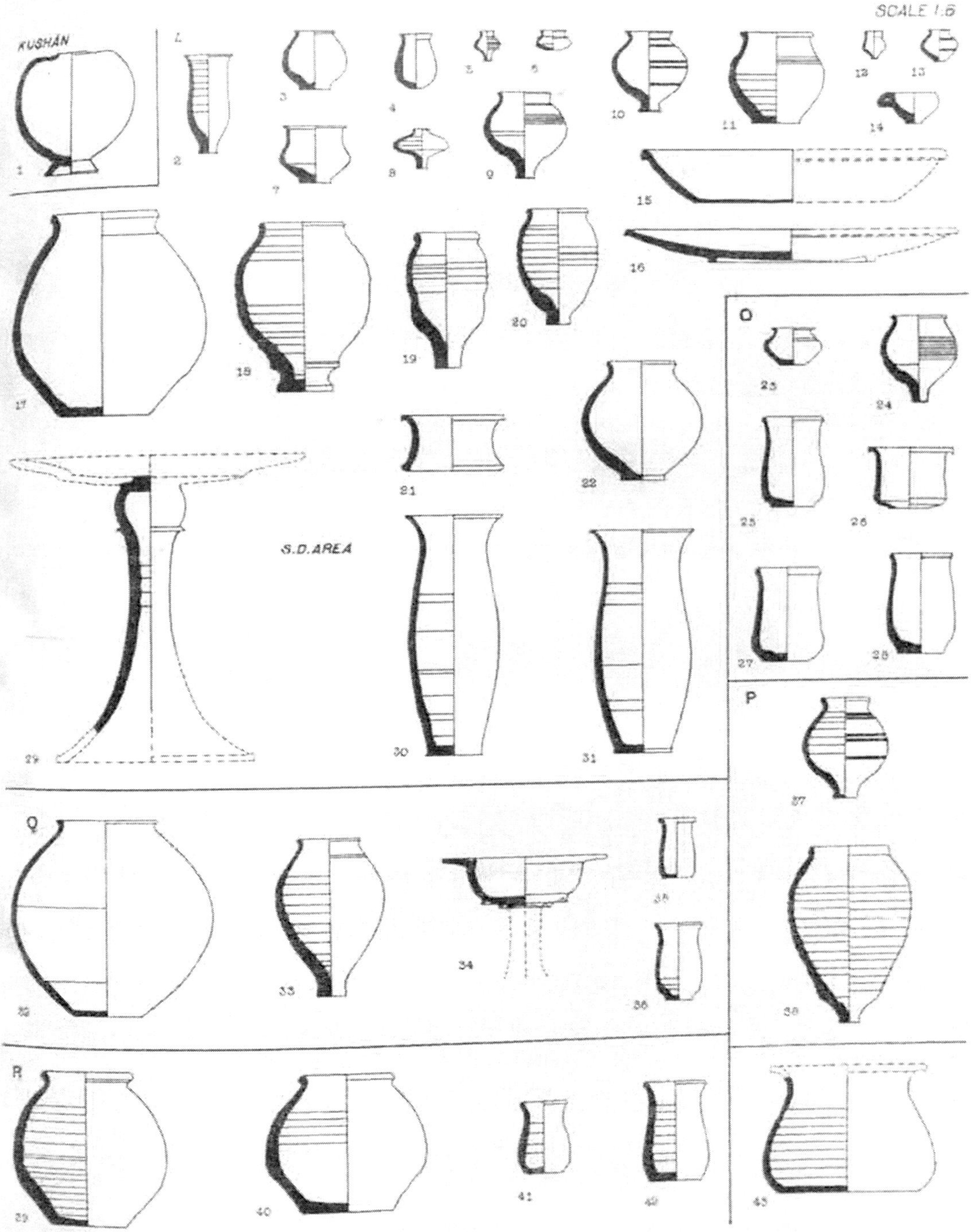

MOHENJO-DARO. SD. AREA. G. SECTION.
POTTERY GROUPS AND GAMESMEN. UPPER LEVELS

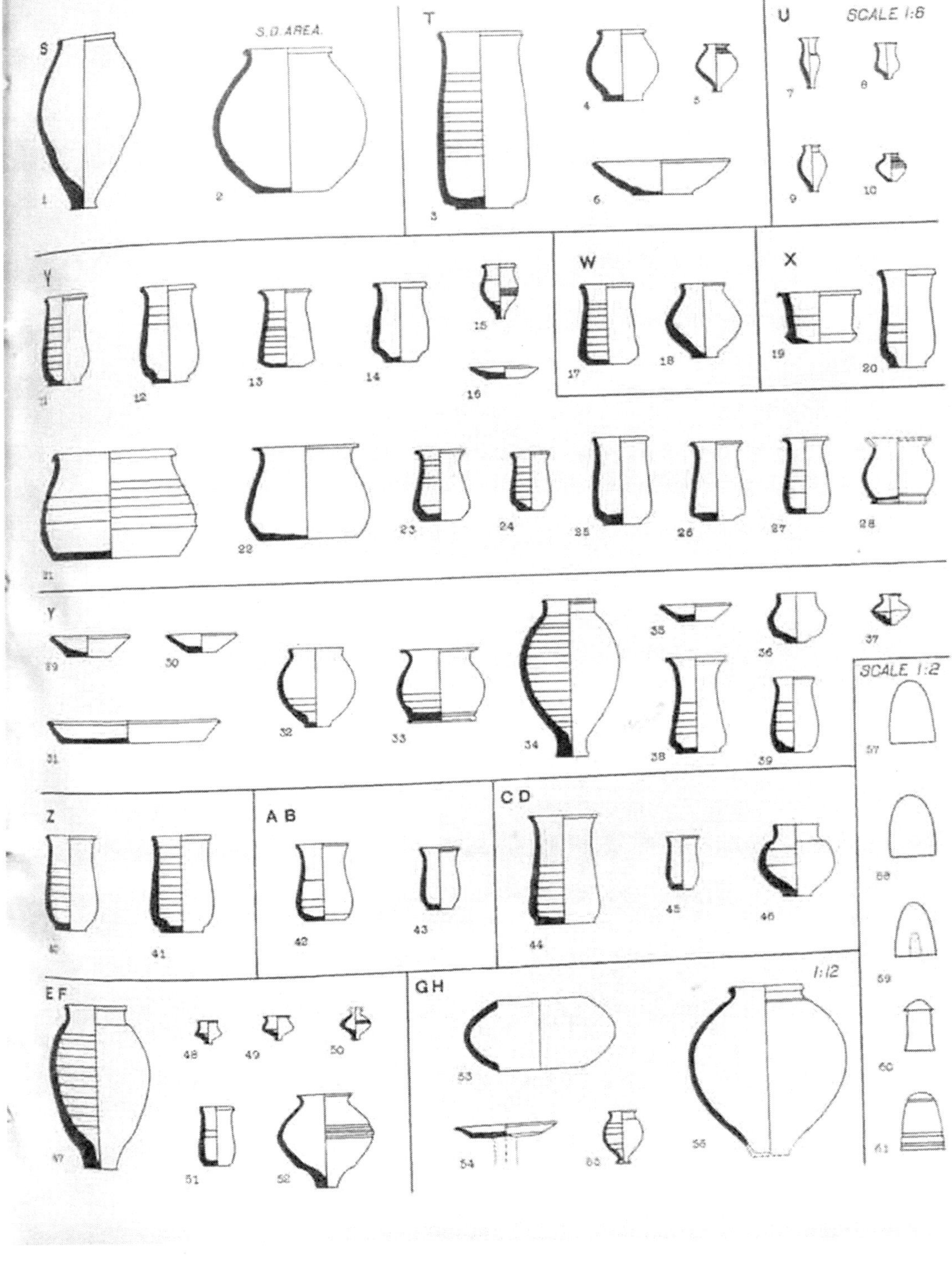

MOHENJO-DARO, DK. AREA, G. SECTION,
POTTERY AND OTHER OBJECTS

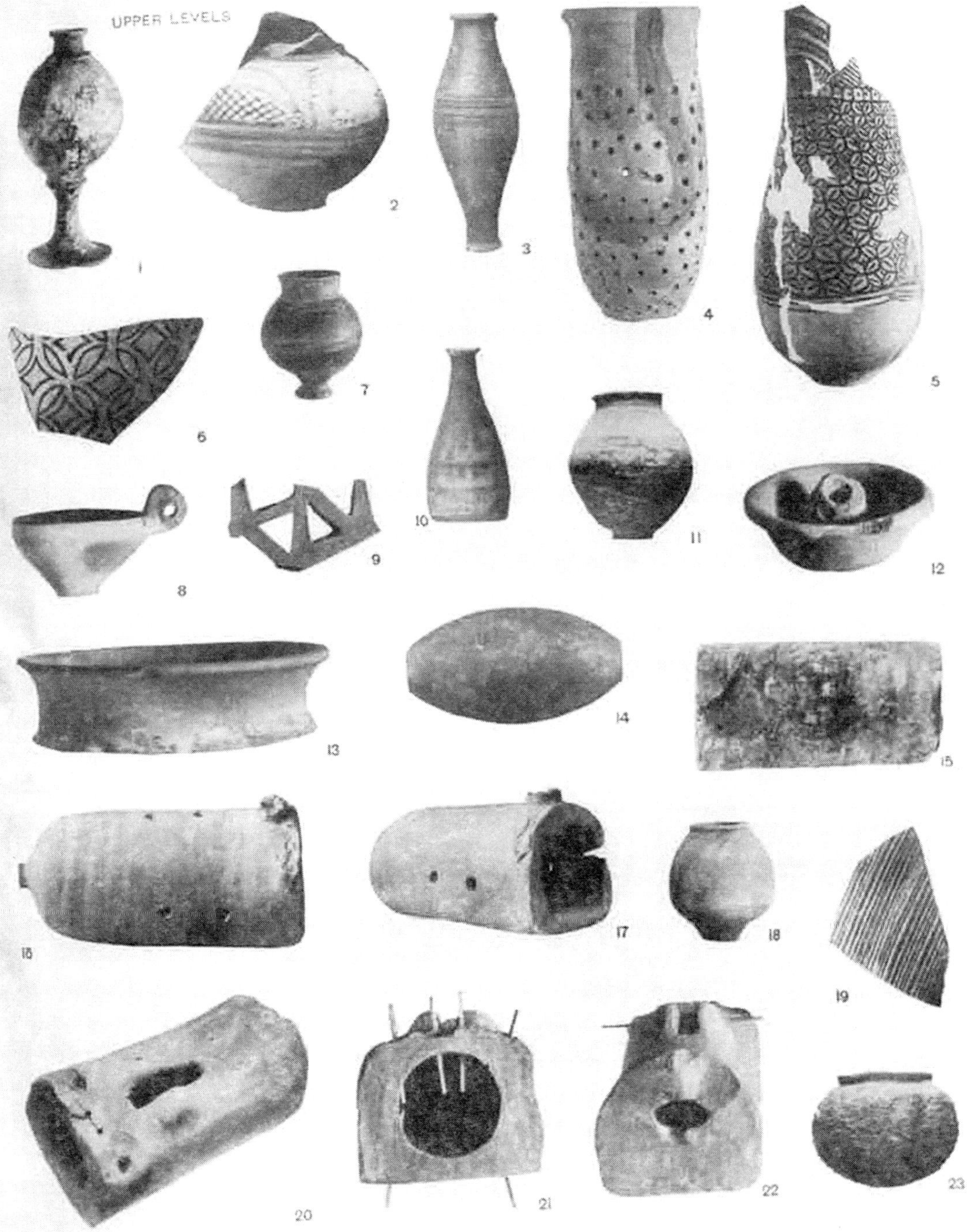

MOHENJO-DARO. DK. AREA, G. SECTION.
POTTERY.

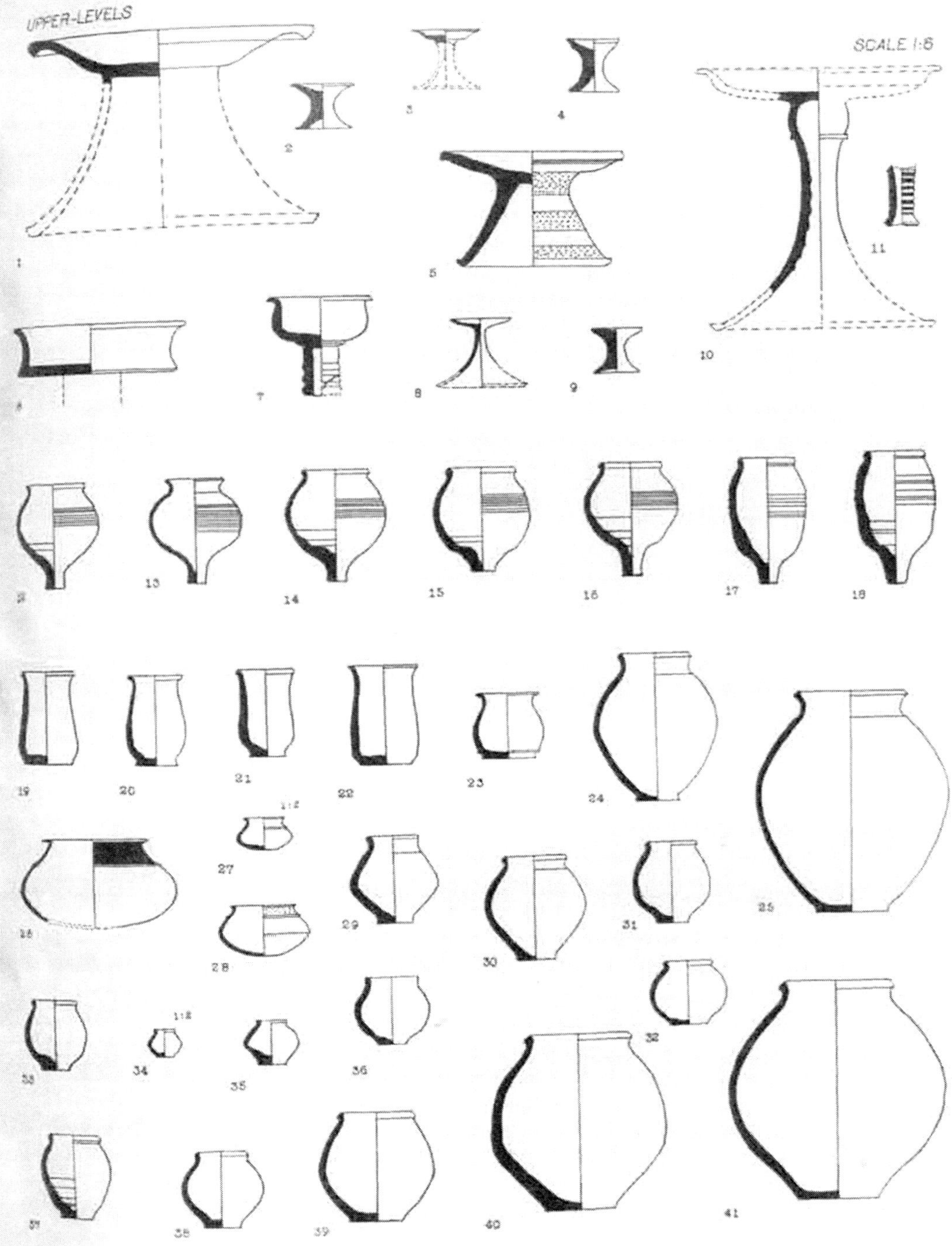

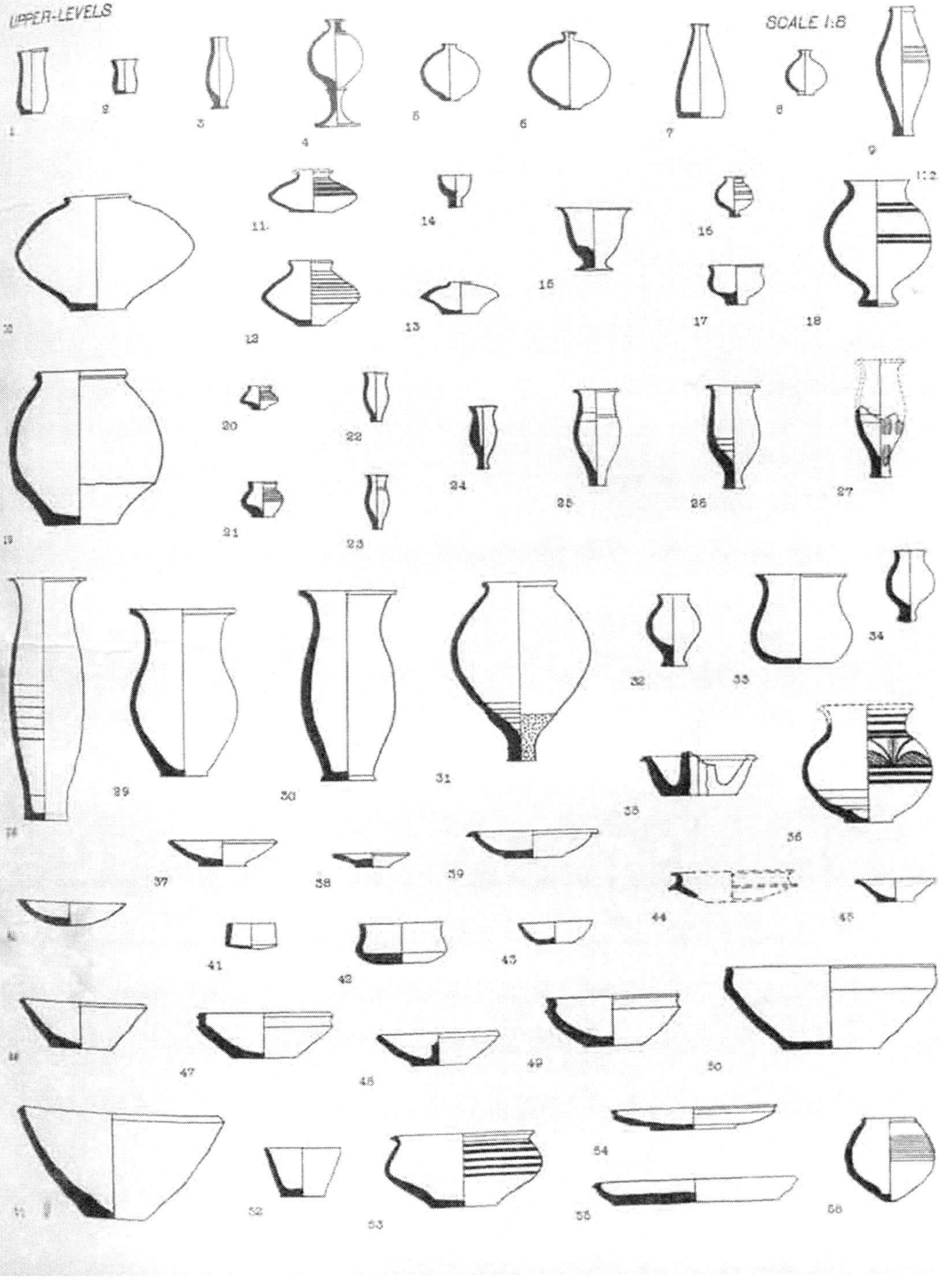
UPPER-LEVELS
SCALE 1:8

MOHENJO-DARO, DK. AREA, G. SECTION.
POTTERY.

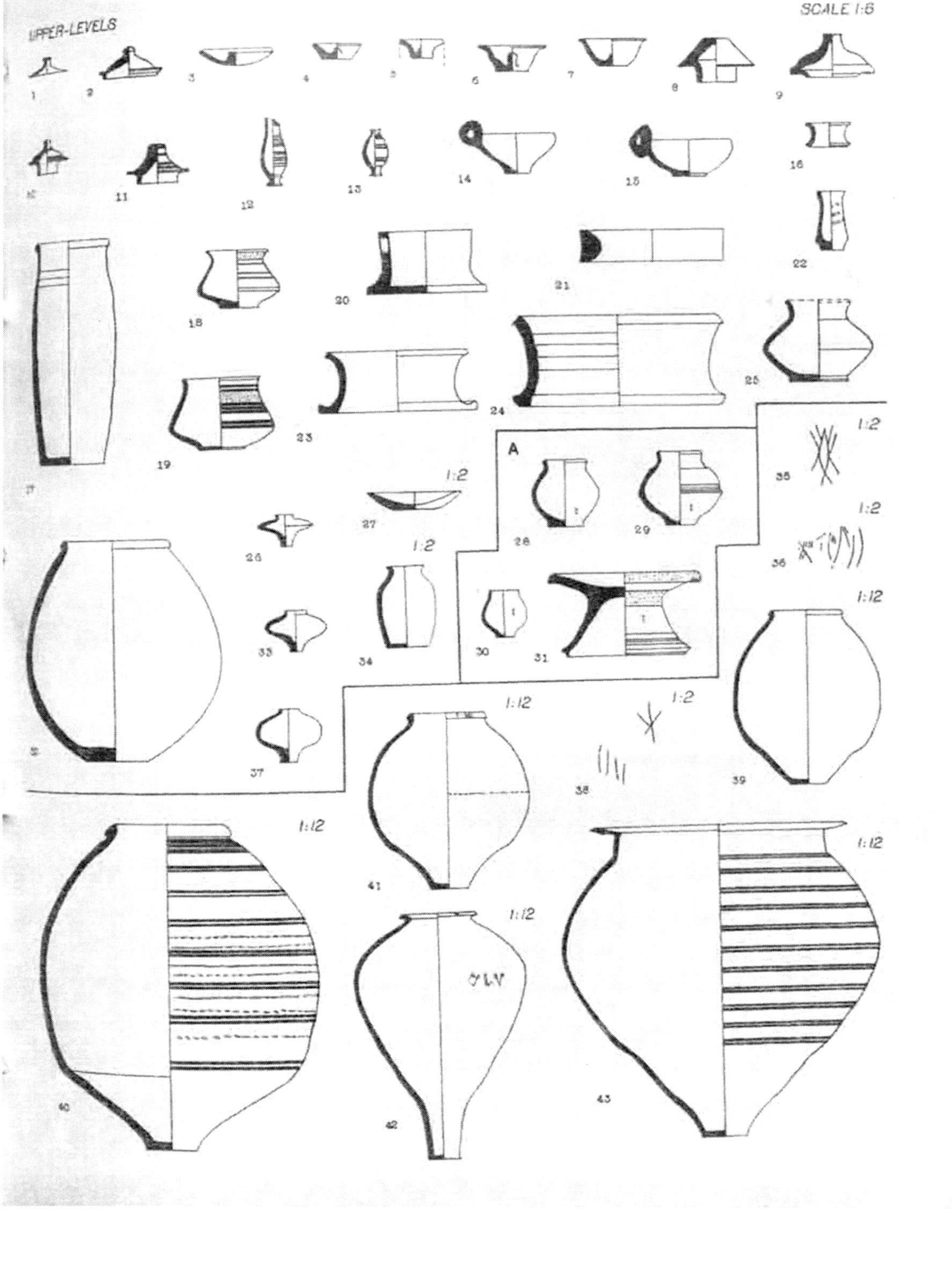

MOHENJO-DARO, DK. AREA. G. SECTION.
STORAGE JARS.

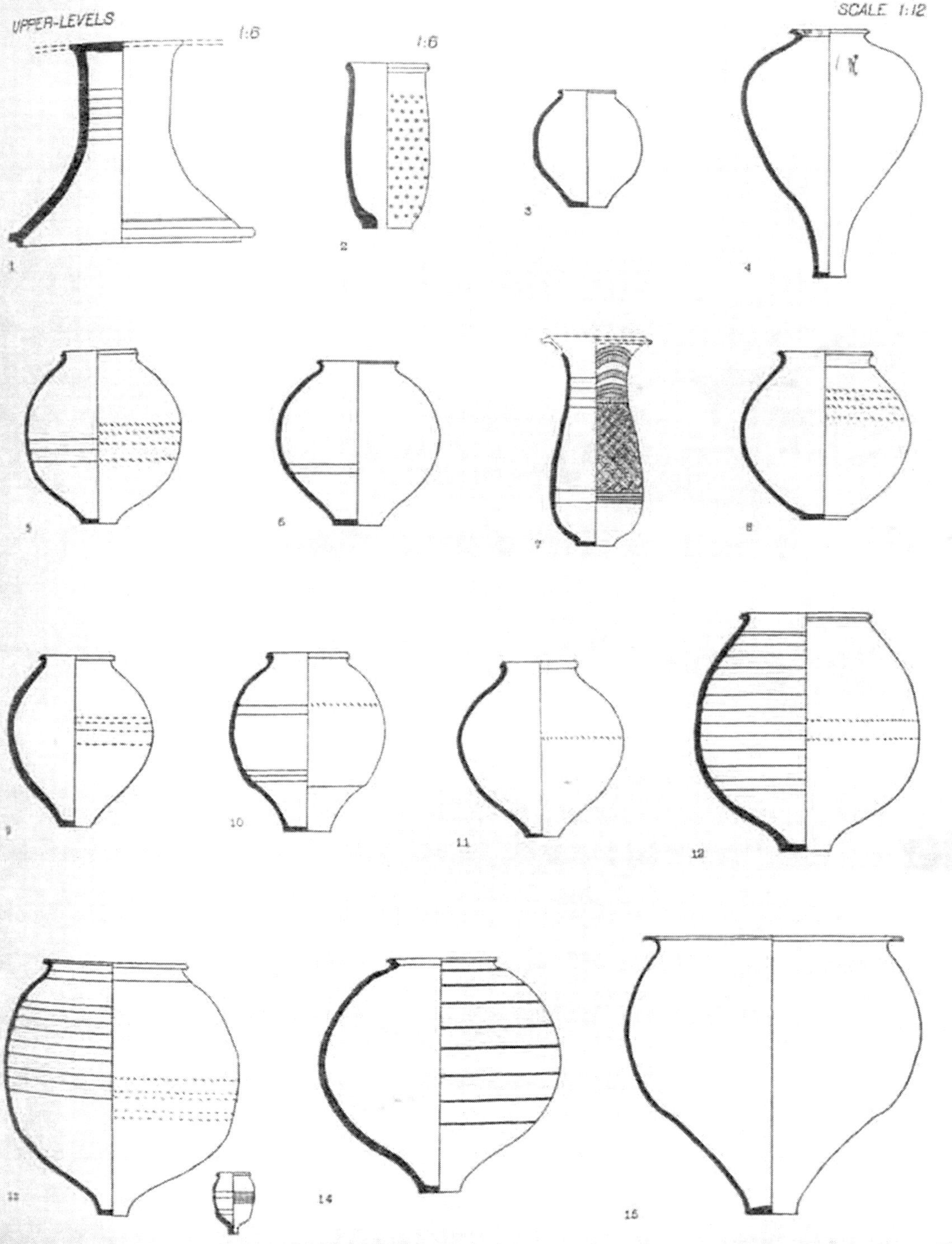

MOHENJO-DARO, DK. AREA, G. SECTION.
POTTERY GROUPS.

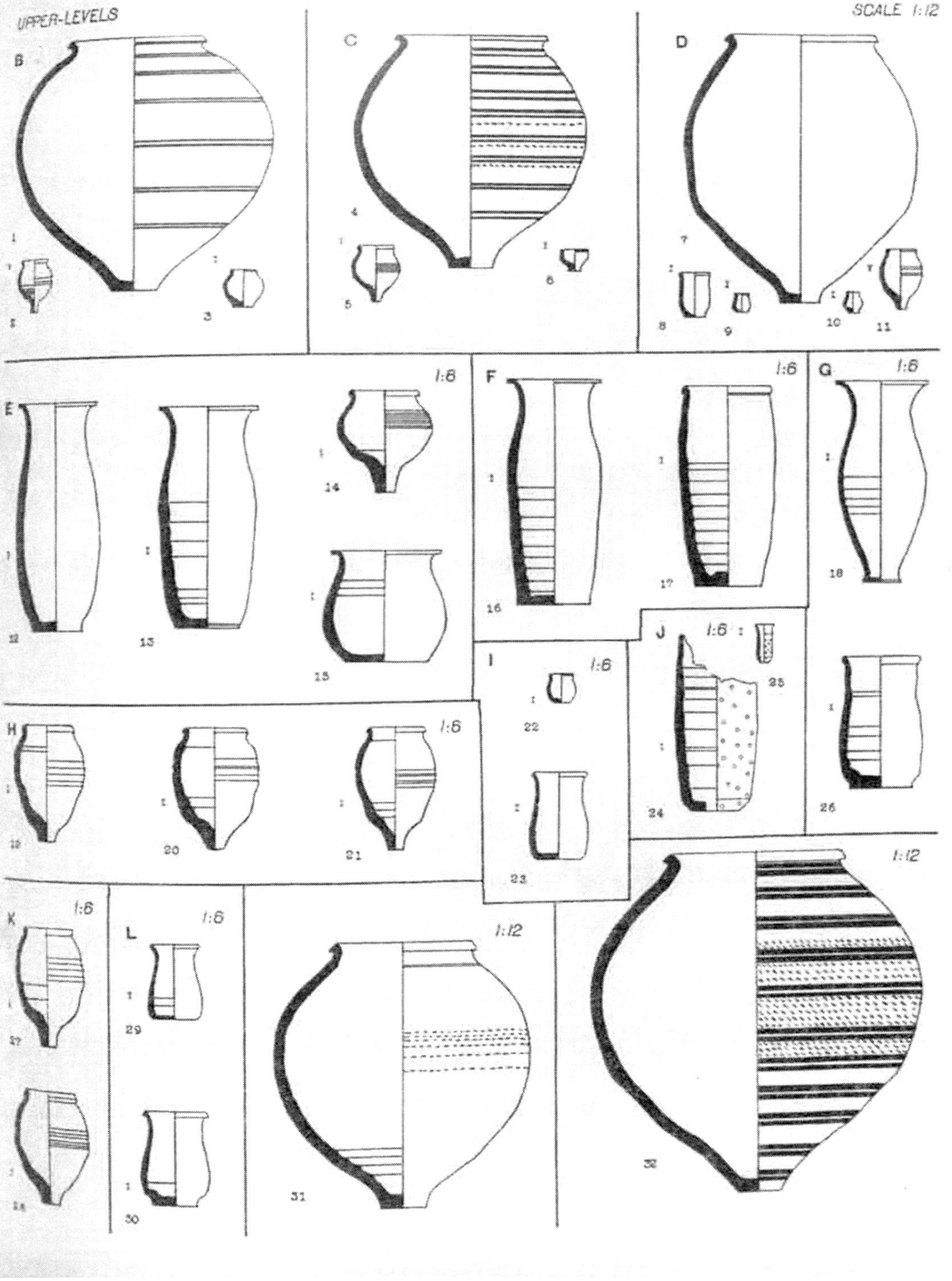

MOHENJO-DARO, DK. AREA, G. SECTION.

POTTERY.

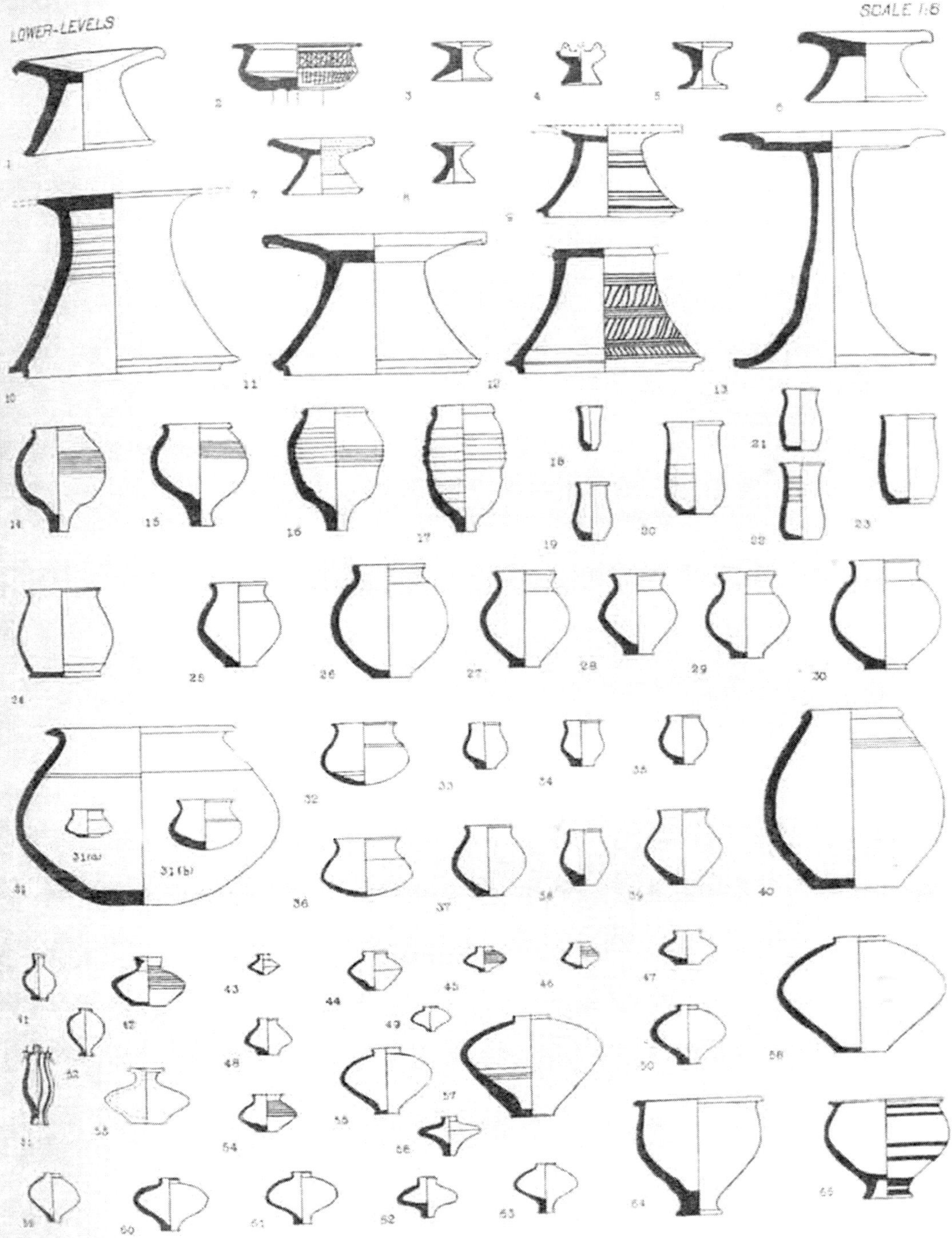

POTTERY.

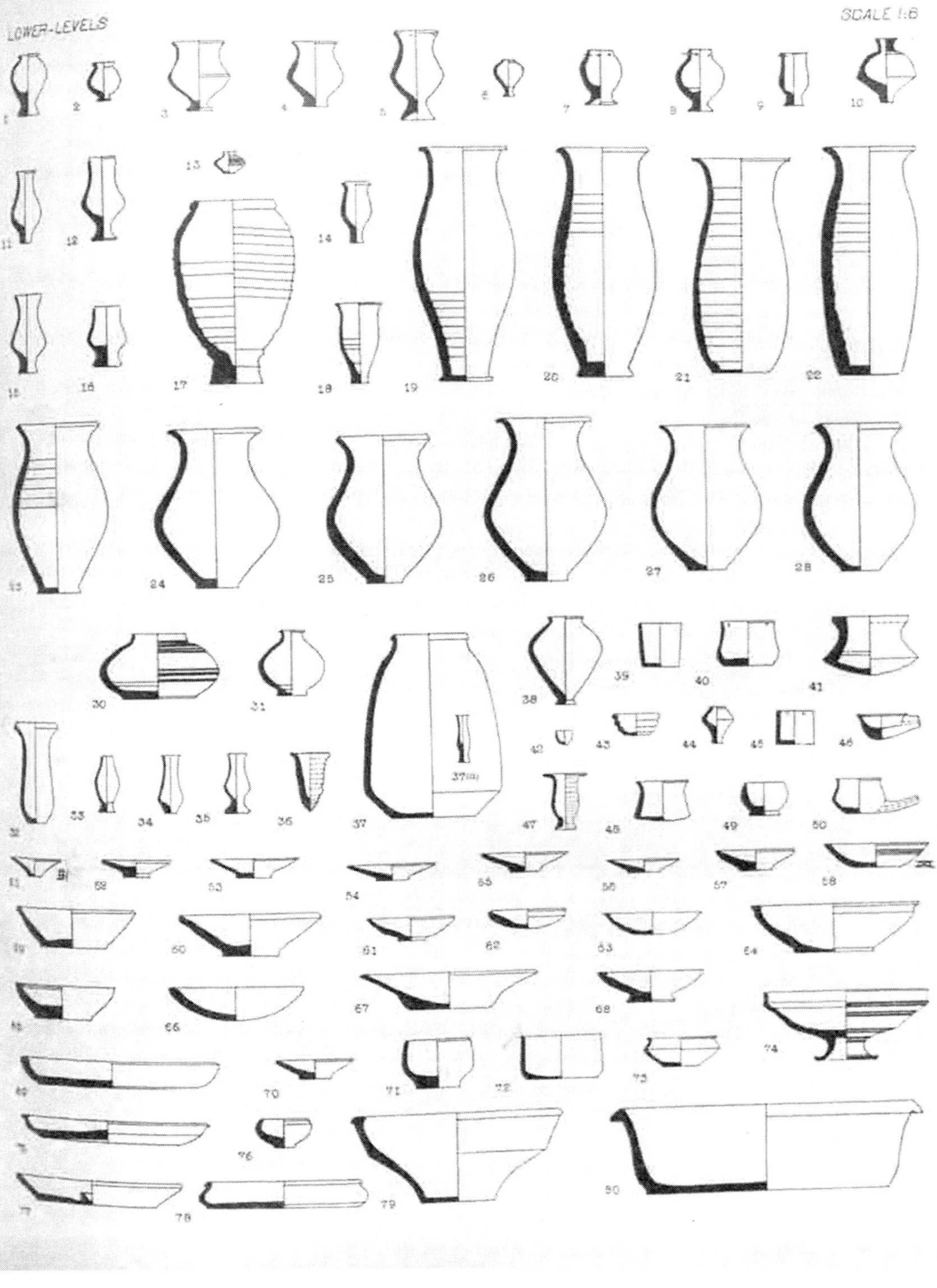

MOHENJO-DARO, DK. AREA, G. SECTION.
POTTERY.

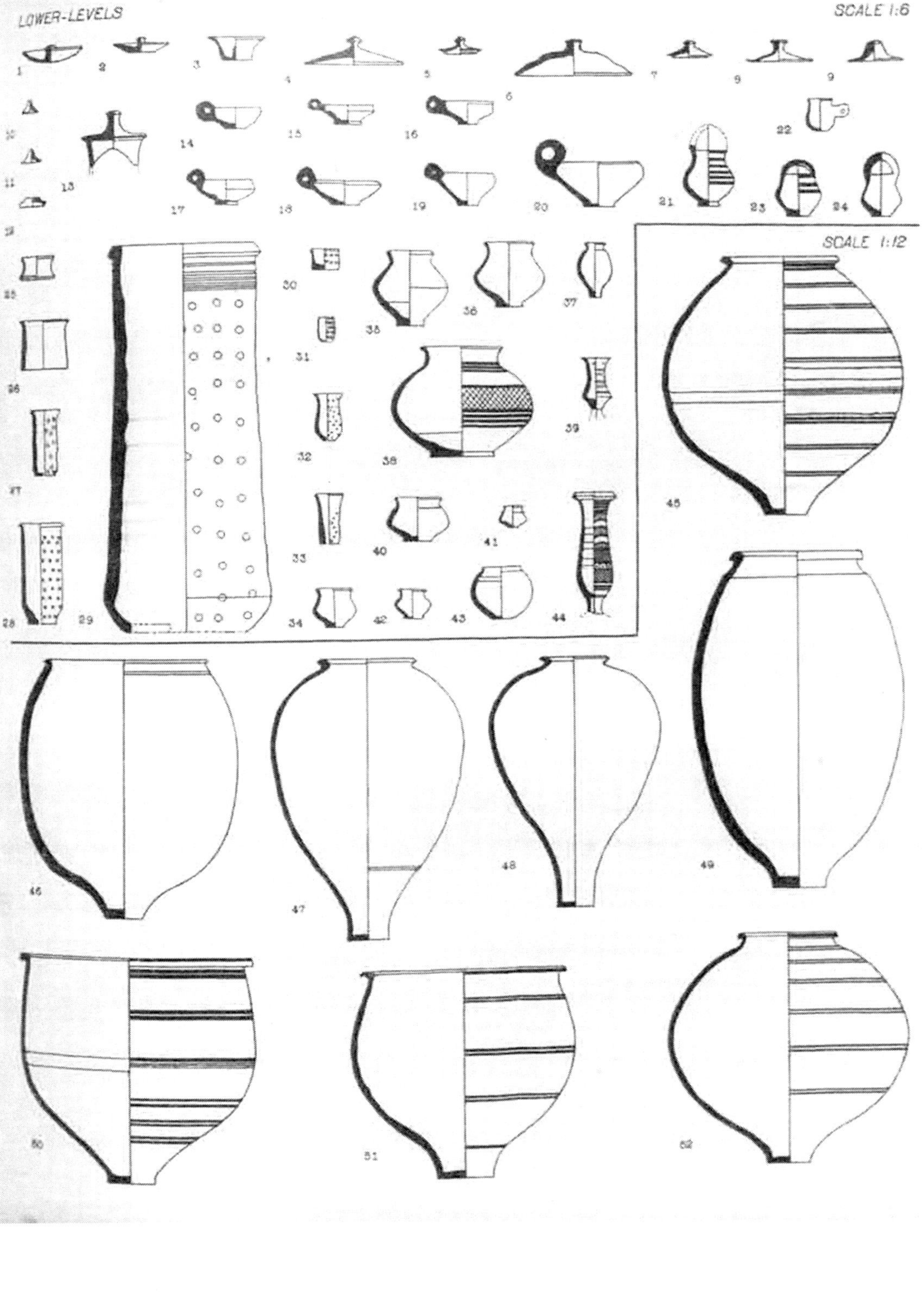

POTTERY STORAGE JARS.

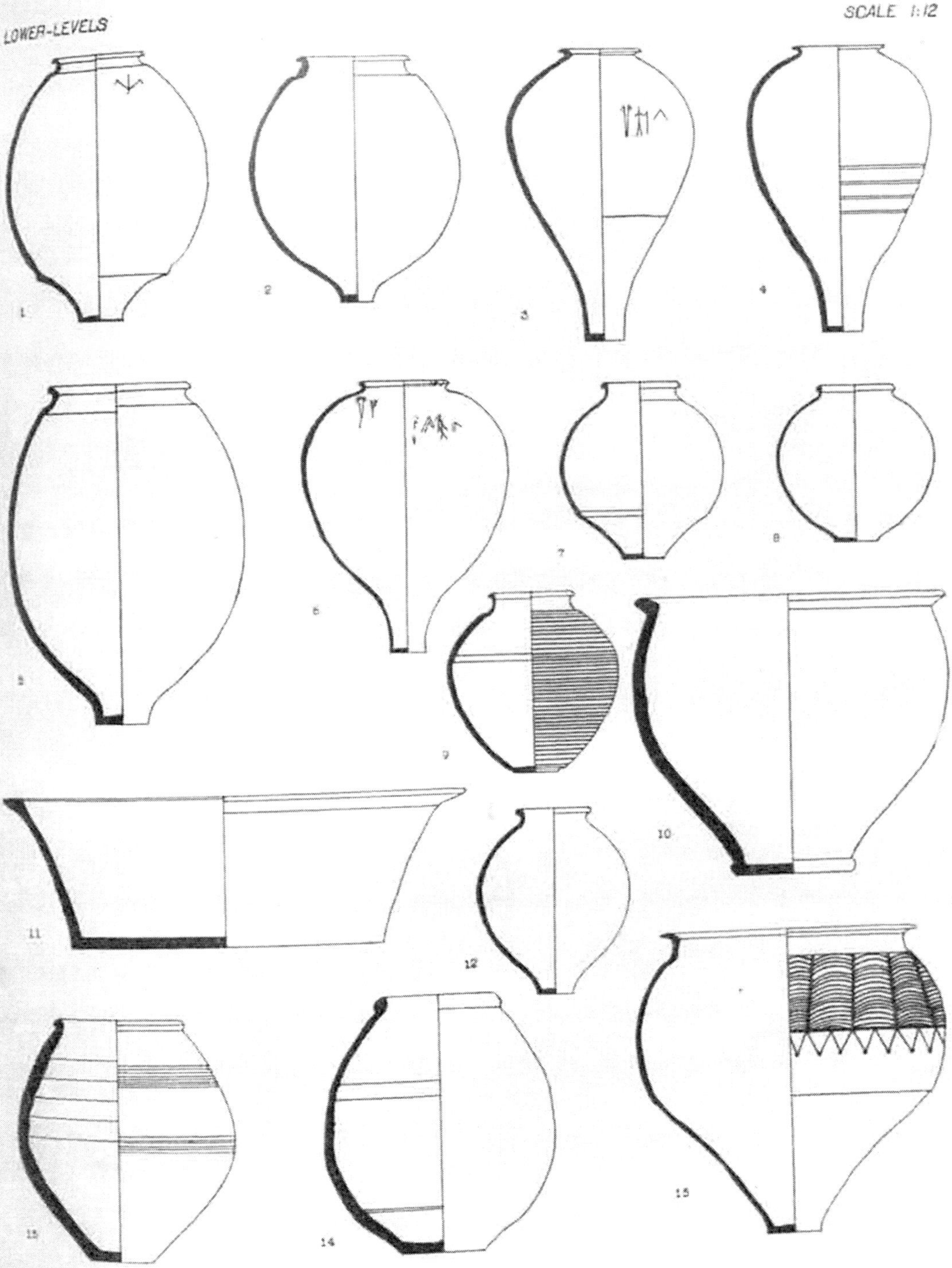

POTTERY GROUPS.

SCALE 1:6

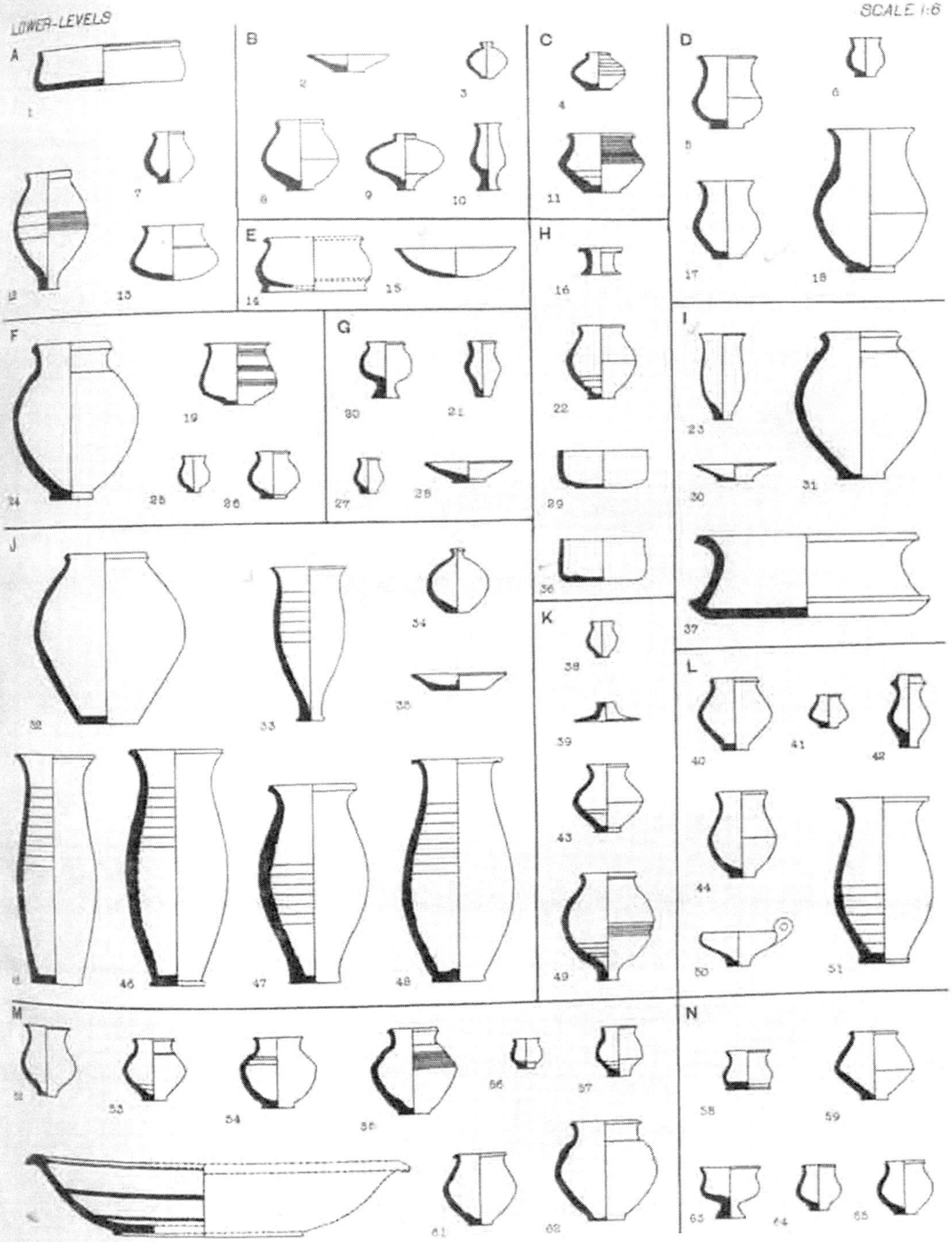

POTTERY GROUPS

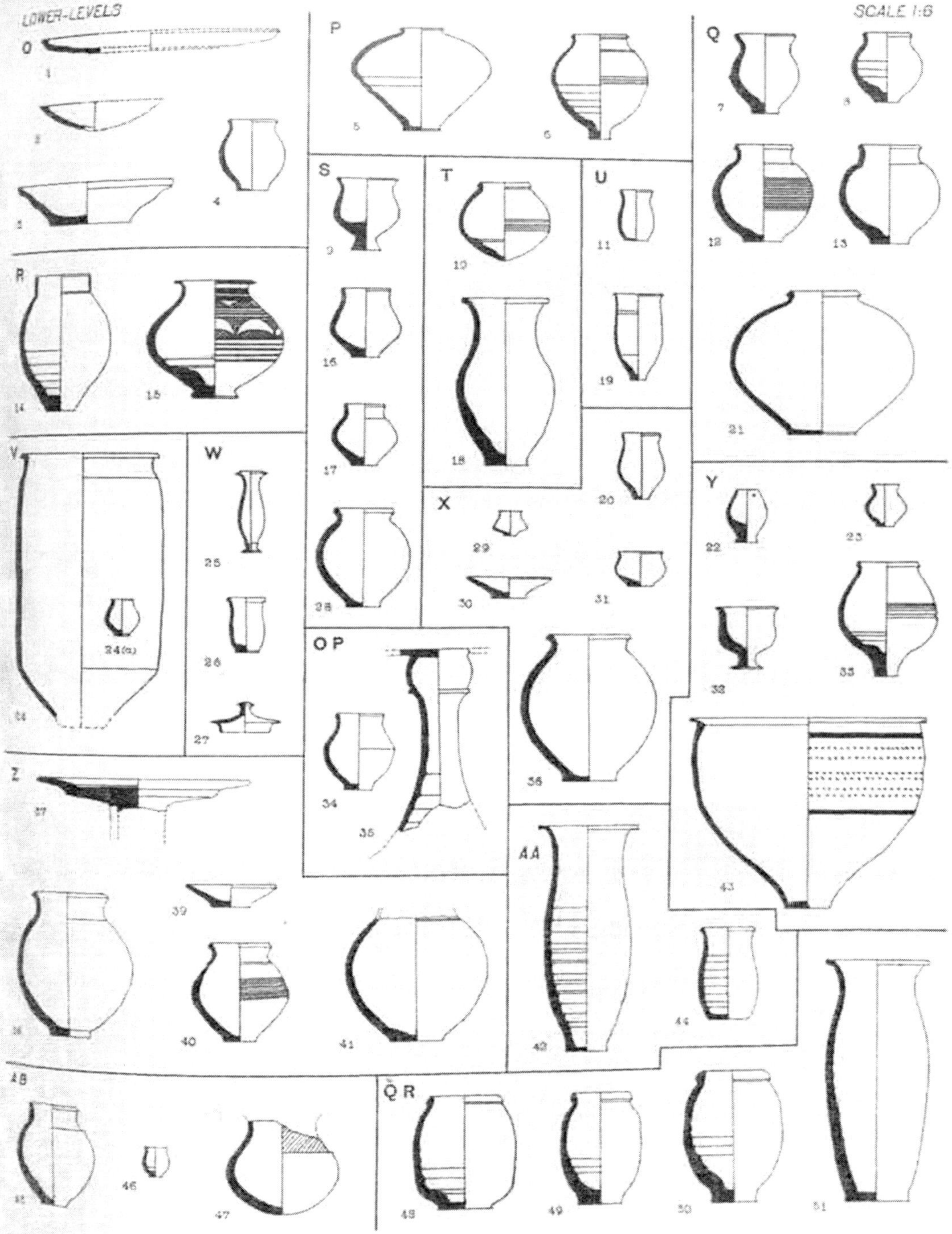

MOHENJO-DARO. DK. AREA. G. SECTION.
POTTERY.

MOHENJO-DARO, DK. AREA, G SECTION.
CUT AND INCISED POTTERY FROM LOWER LEVELS.

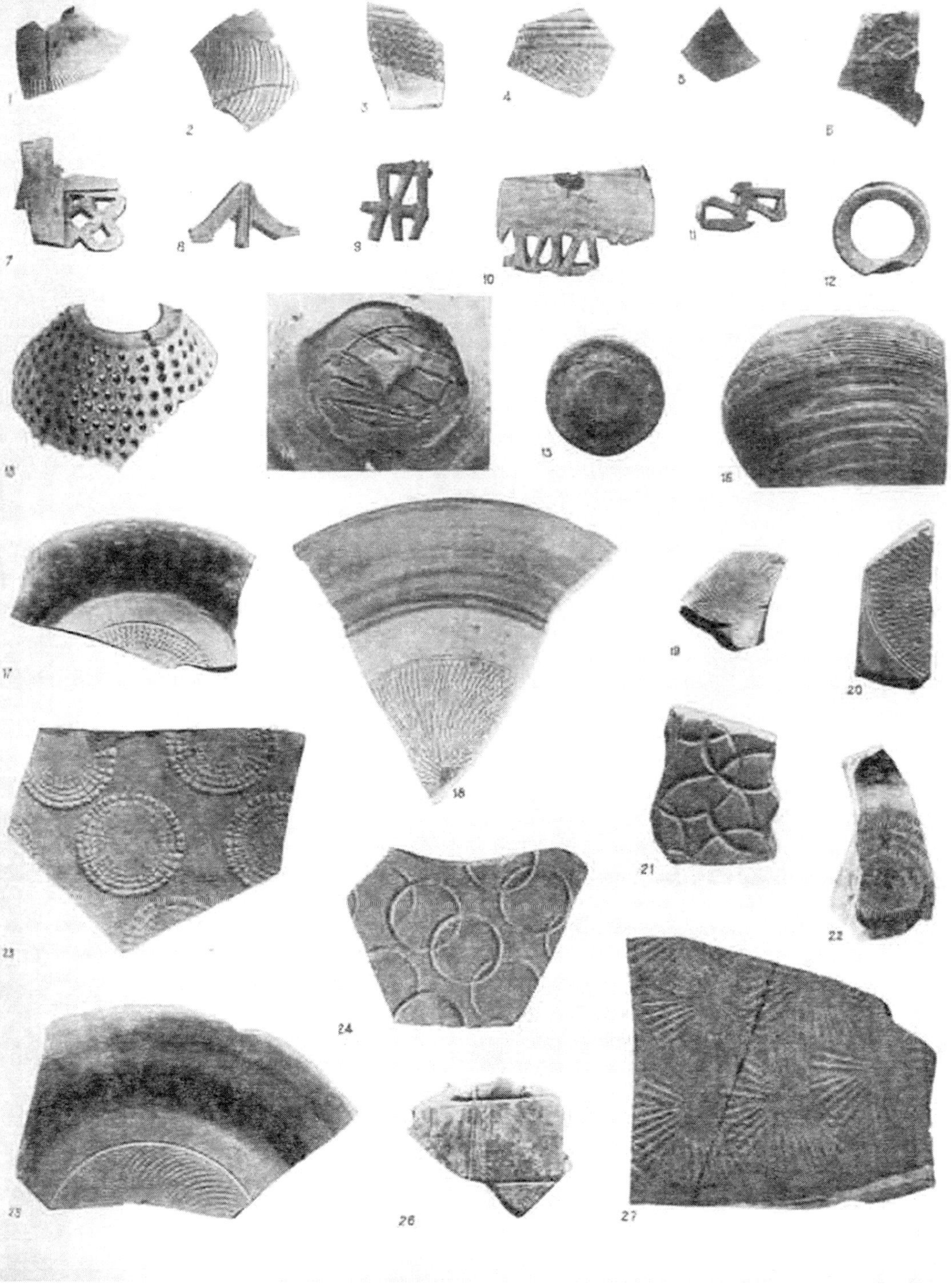

MOHENJO-DARO, DK. AREA, G. SECTION.
FRAGMENTS OF PAINTED POTTERY.

UPPER LEVELS

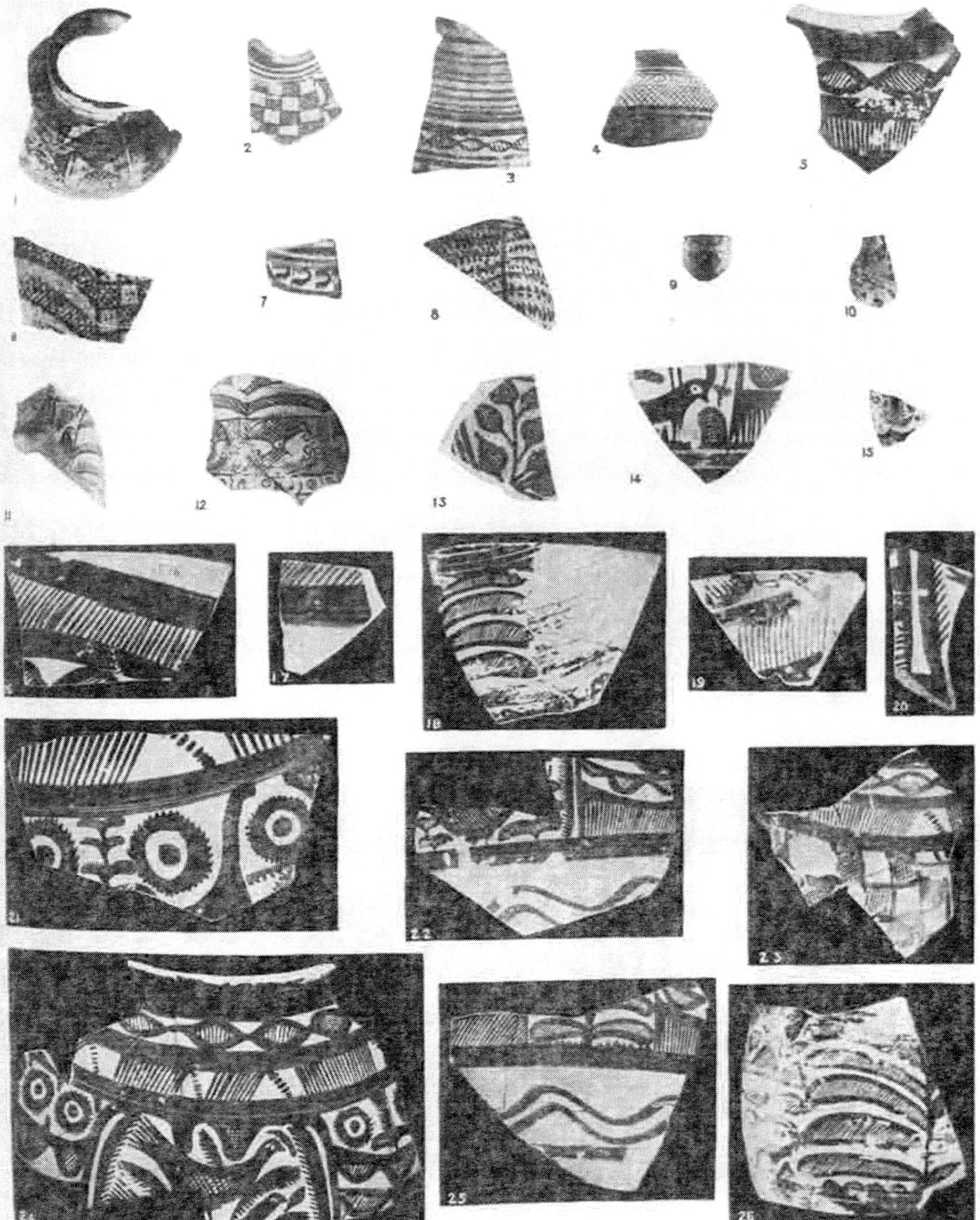

PAINTED POTTERY MOTIFS.

UPPER LEVELS

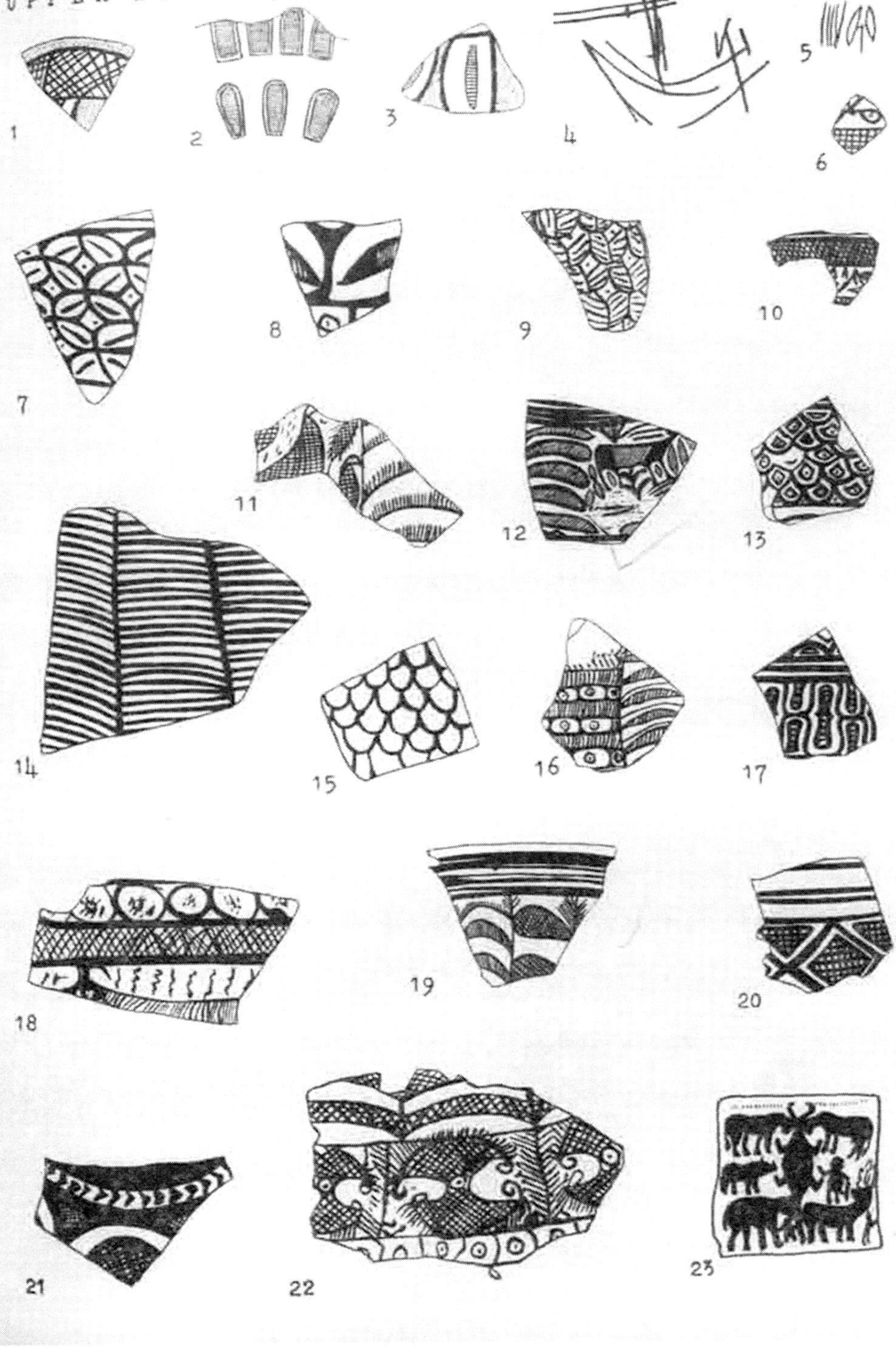

FRAGMENTS OF PAINTED POTTERY.

MOHENJO-DARO. SD. AREA.
MISCELLANEOUS OBJECTS.

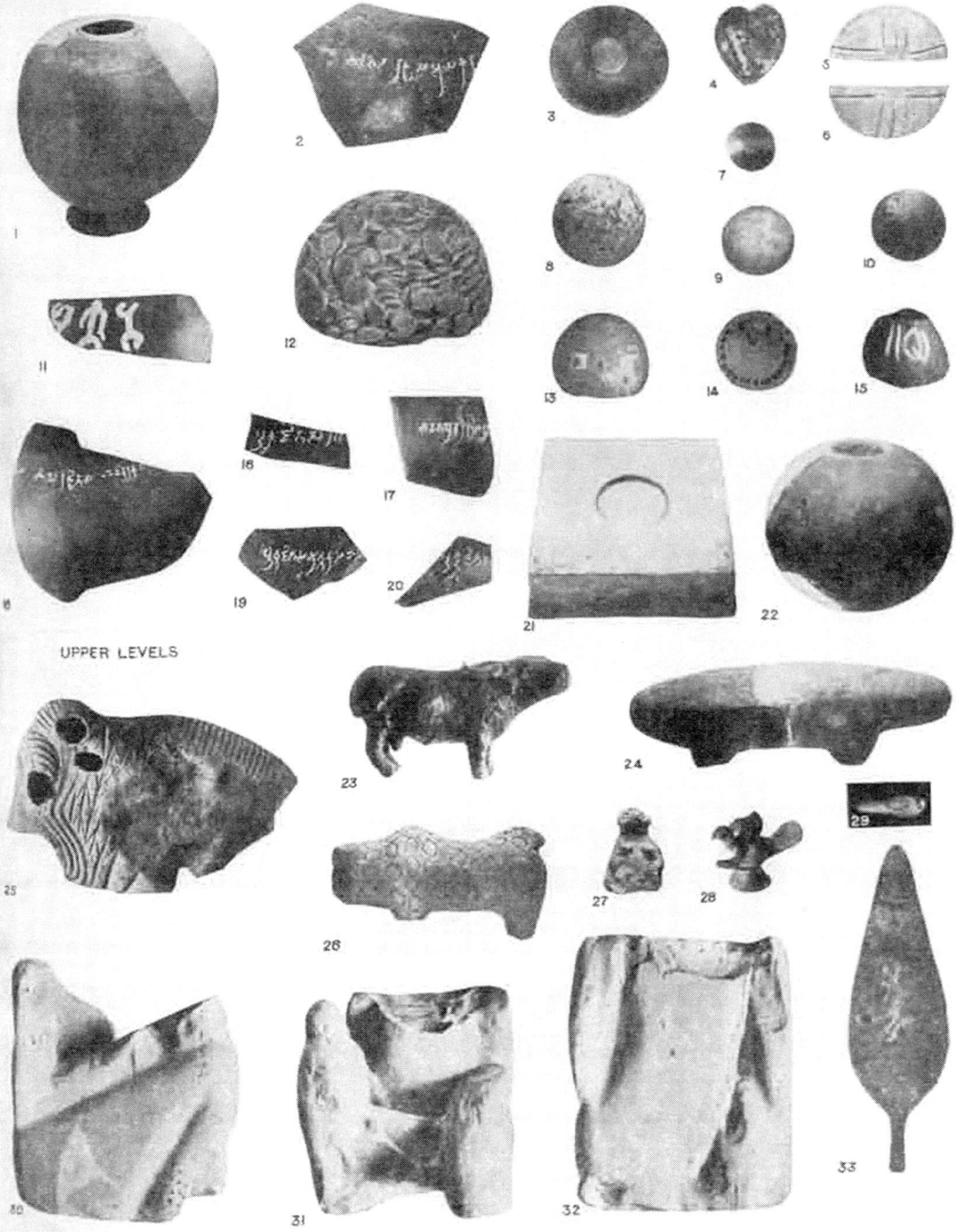

MOHENJO-DARO, DK. AREA, G. SECTION.
POTTERY FIGURINES. UPPER LEVELS.

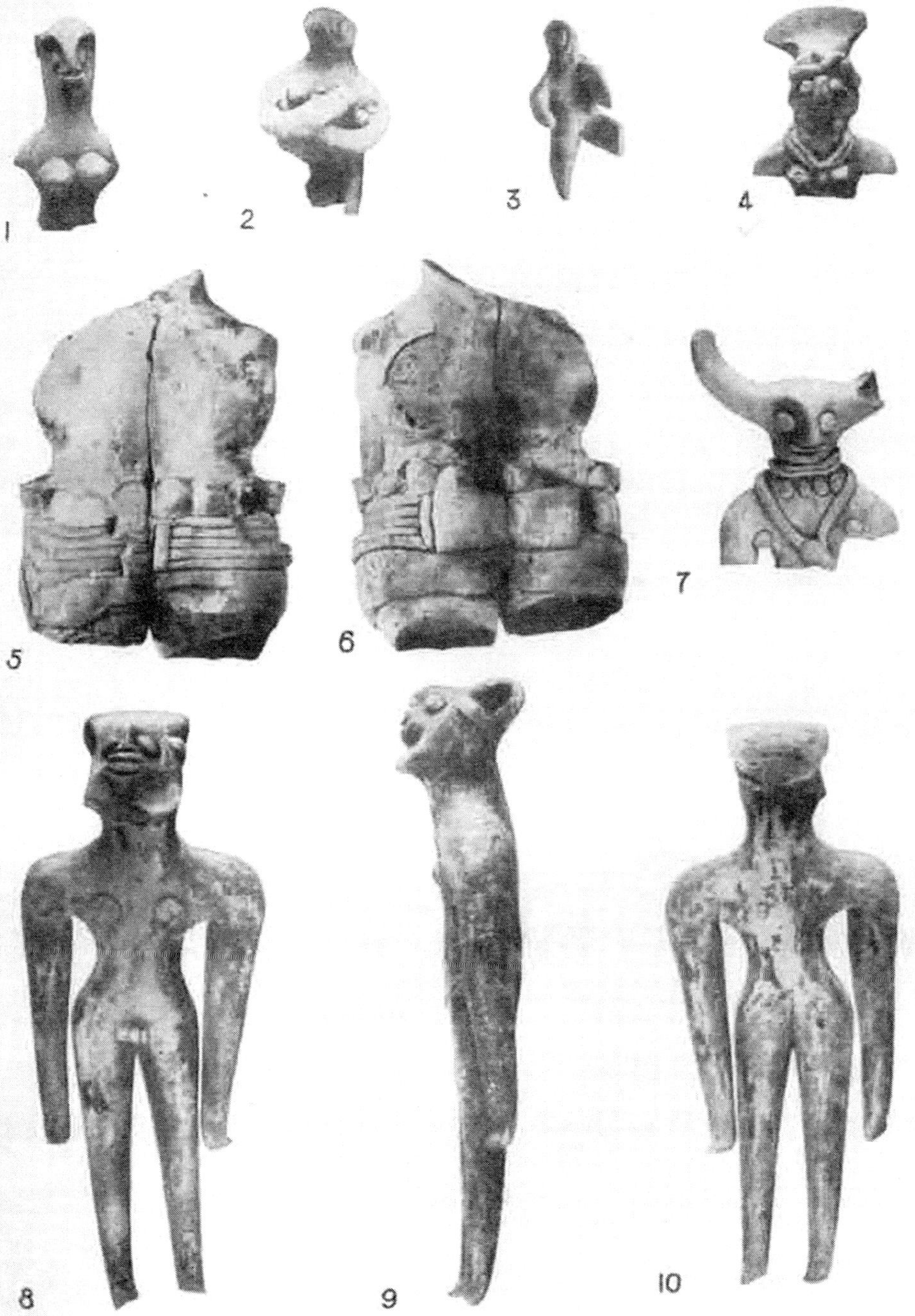

MOHENJO-DARO, DK. AREA, G. SECTION.
POTTERY AND BRONZE FIGURINES. UPPER LEVELS.

MOHENJO-DARO, DK. AREA, G. SECTION.

POTTERY, SHELL, FAIENCE AND BRONZE MODEL ANIMALS.

MOHENJO-DARO, DK. AREA, G. SECTION.

POTTERY FIGURINES. LOWER LEVELS.

MOHENJO-DARO: DK. AREA, G. SECTION.
POTTERY AND PASTE FIGURINES. LOWER LEVELS.

MOHENJO-DARO, DK. AREA, G. SECTION.
POTTERY, STONE, FAIENCE AND BRONZE MODEL ANIMALS. UPPER LEVELS.

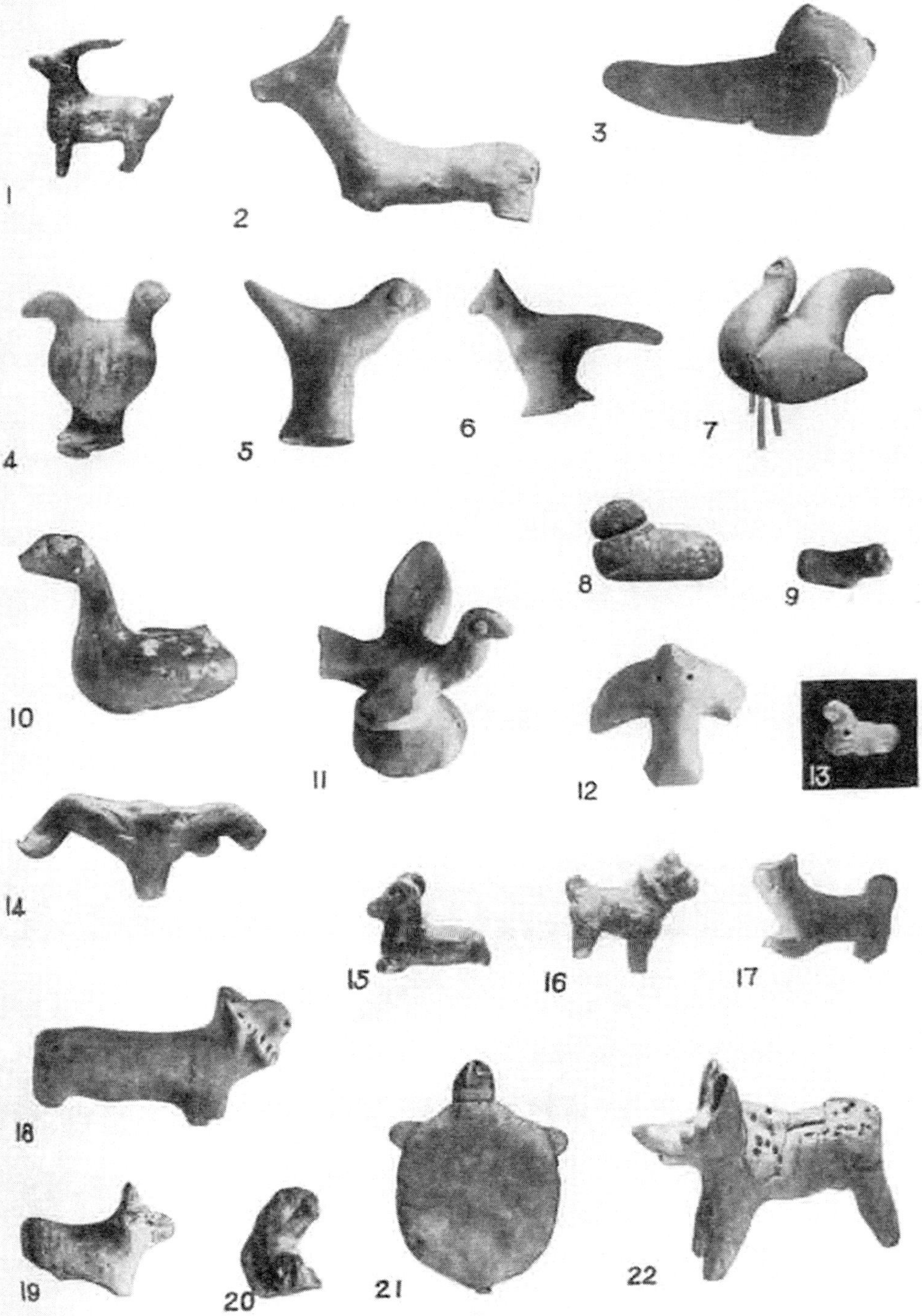

MOHENJO-DARO, DK. AREA, G. SECTION.

POTTERY AND FAIENCE ANIMALS.

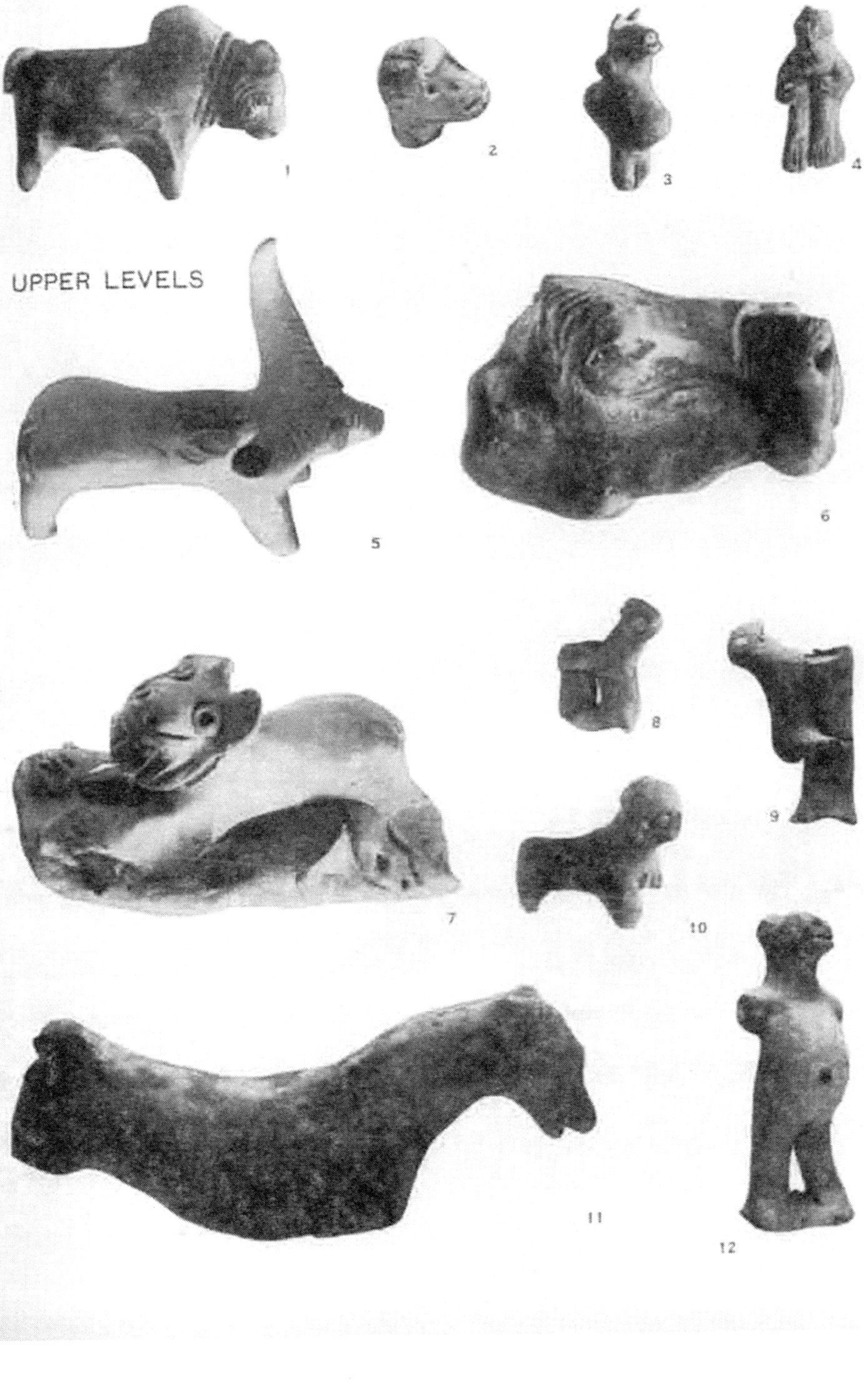

MOHENJO-DARO: DK. AREA, G. SECTION.
POTTERY, STONE, SHELL AND BRONZE MODEL ANIMALS, LOWER LEVELS.

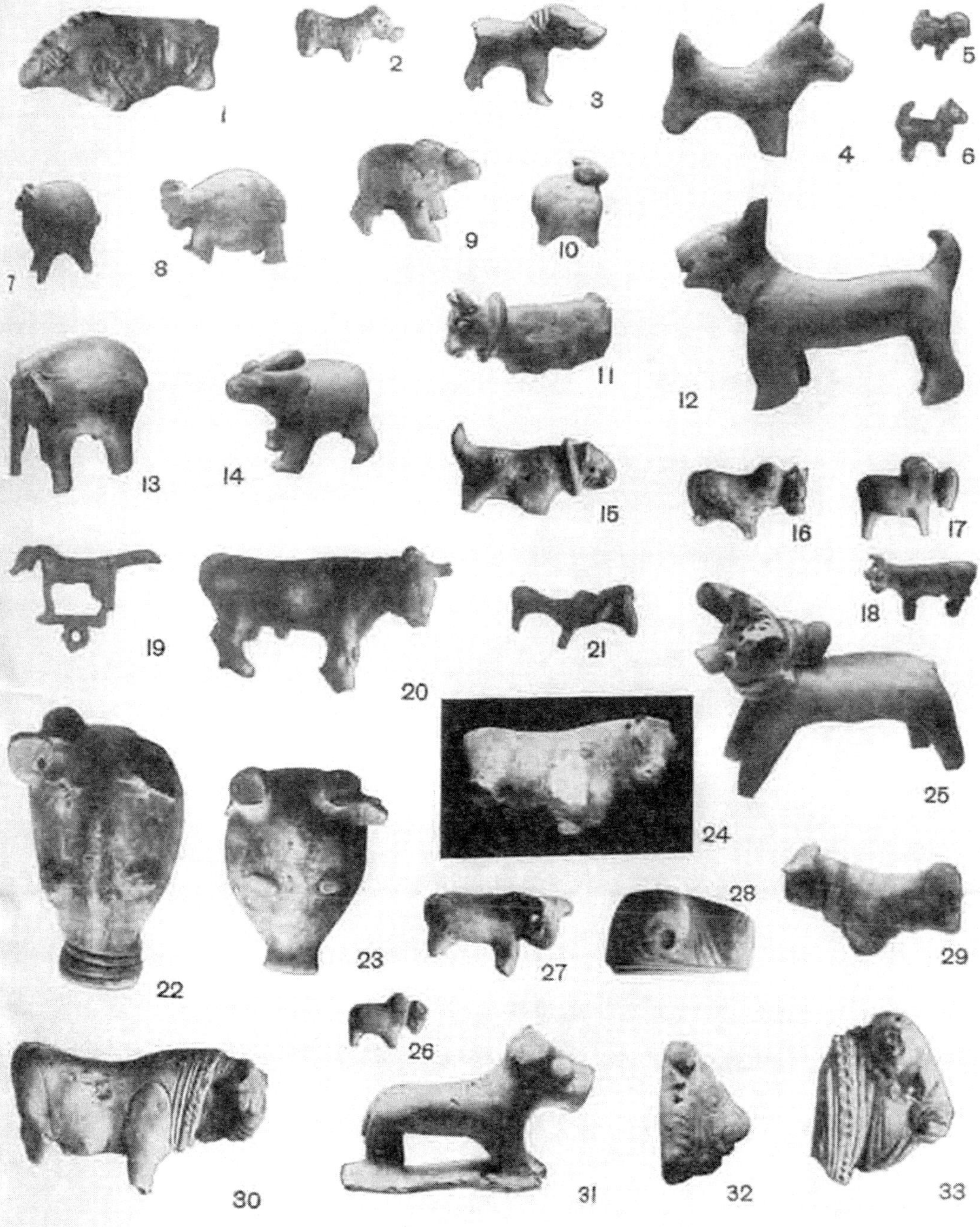

MOHENJO-DARO. DK. AREA. G. SECTION.

POTTERY, PASTE AND FAIENCE MODEL ANIMALS. LOWER LEVELS.

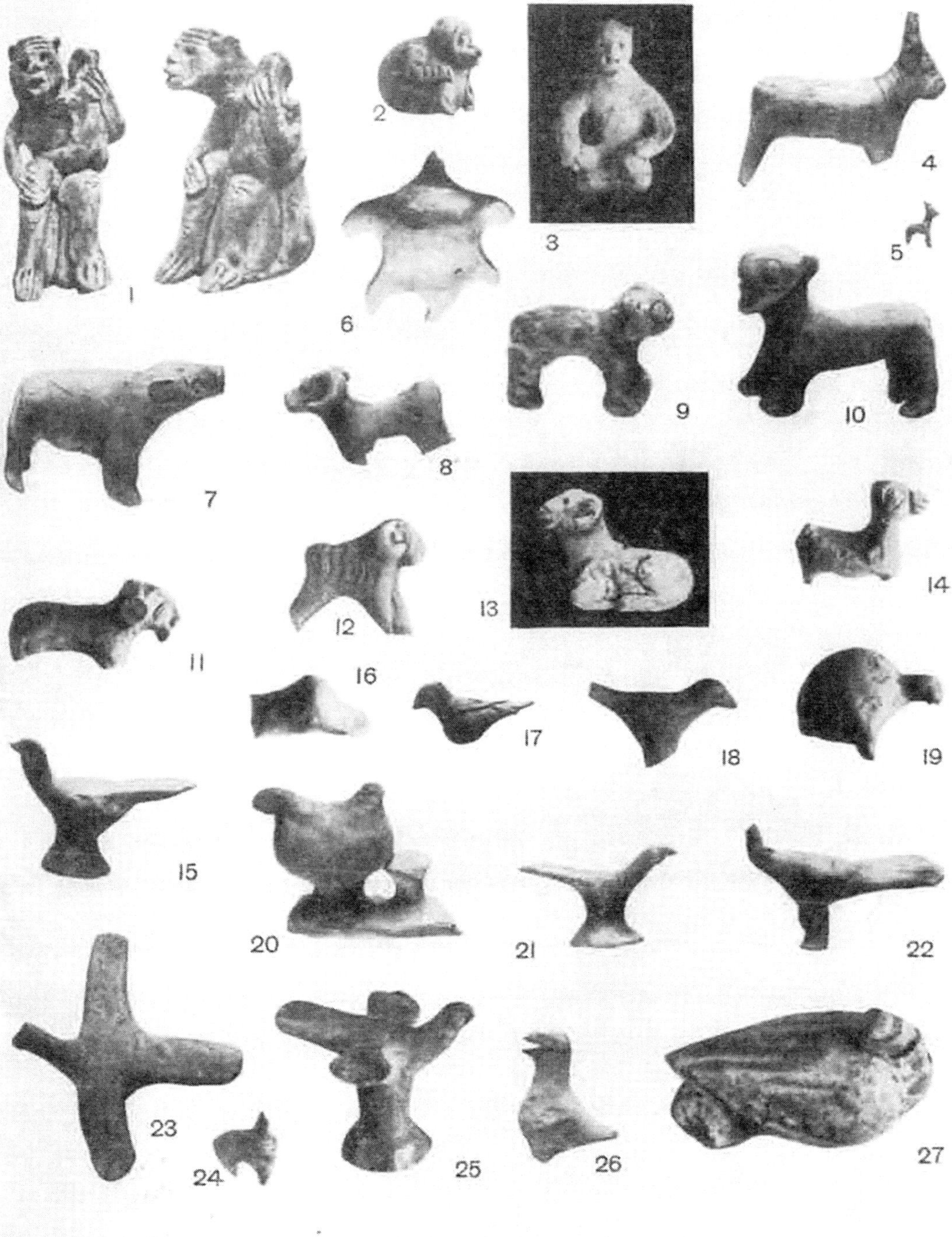

MOHENJO-DARO: DK. AREA. G. SECTION.
POTTERY MODEL ANIMALS, GAMESMEN AND JAR-STOPPERS. LOWER LEVELS.

MOHENJO-DARO : DK. & SD AREAS.
SEALS, AMULETS, BEADS, FIGURINES AND COINS, UPPER LEVELS.

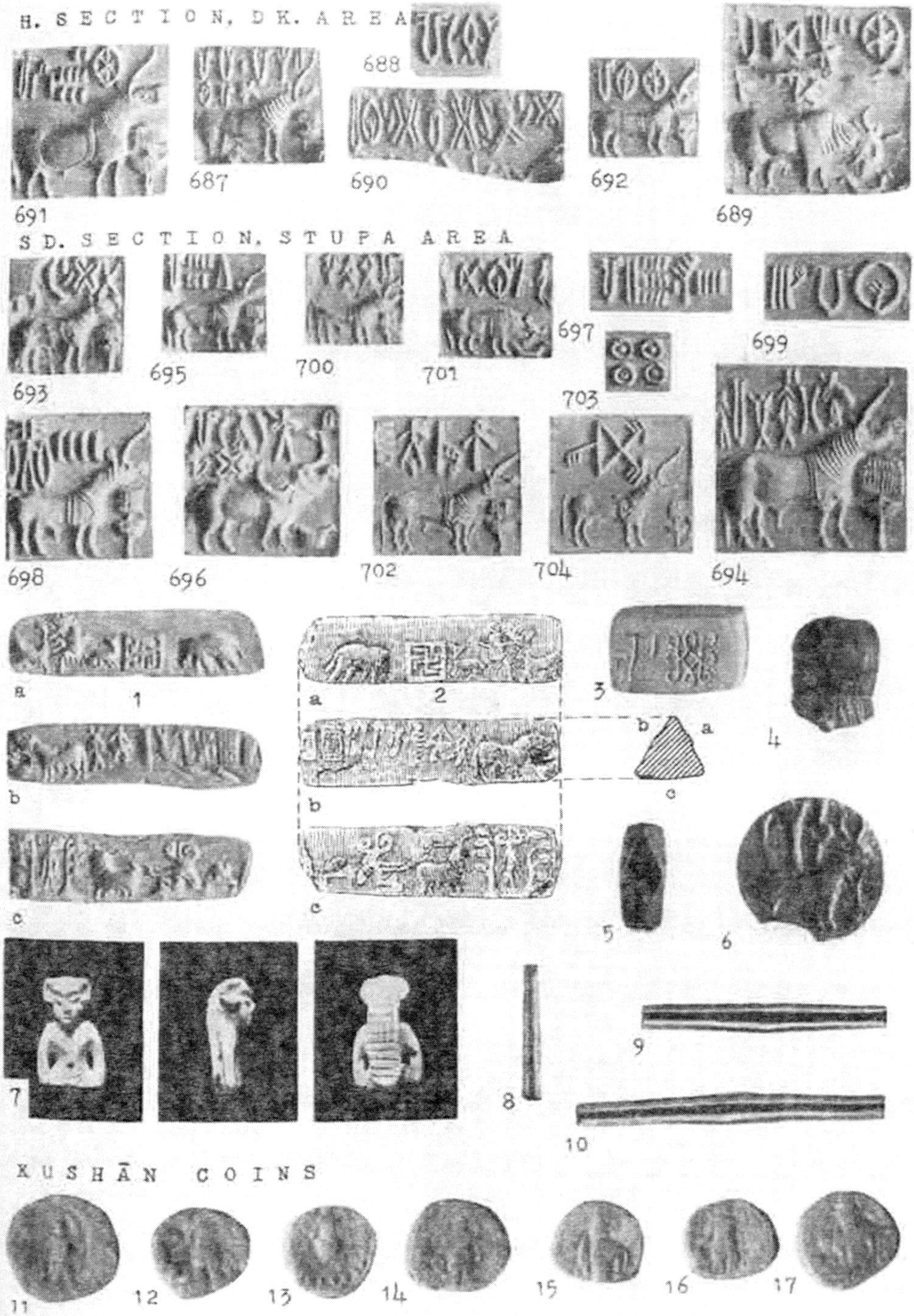

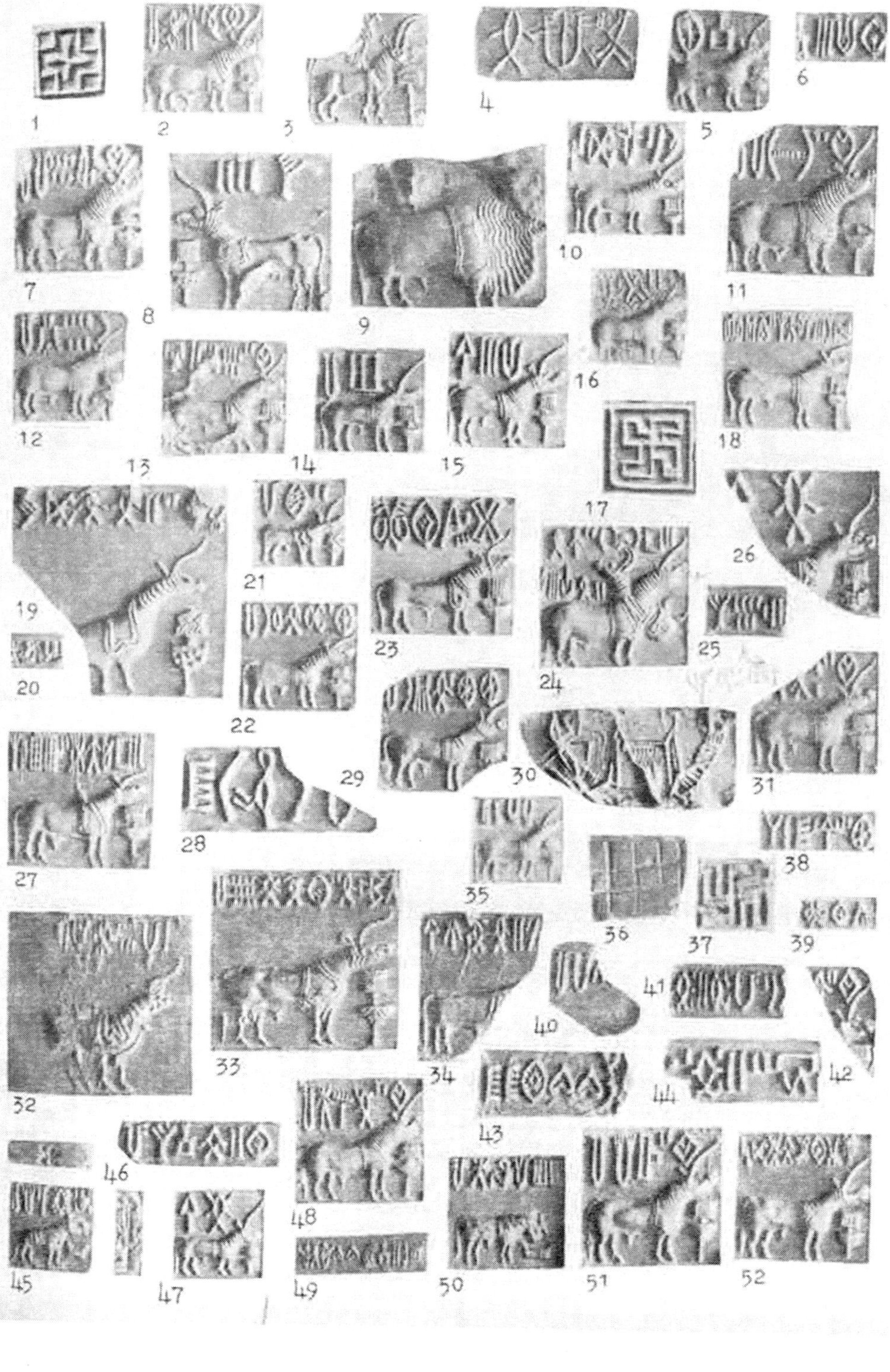
1
2
3
4
5
6
7
8
9
10
11
12
13
14
15
16
17
18
19
20
21
22
23
24
25
26
27
28
29
30
31
32
33
34
35
36
37
38
39
40
41
42
43
44
45
46
47
48
49
50
51
52

MOHENJO-DARO, DK. AREA, G. SECTION.
IMPRESSIONS OF SEALS. UPPER LEVELS.

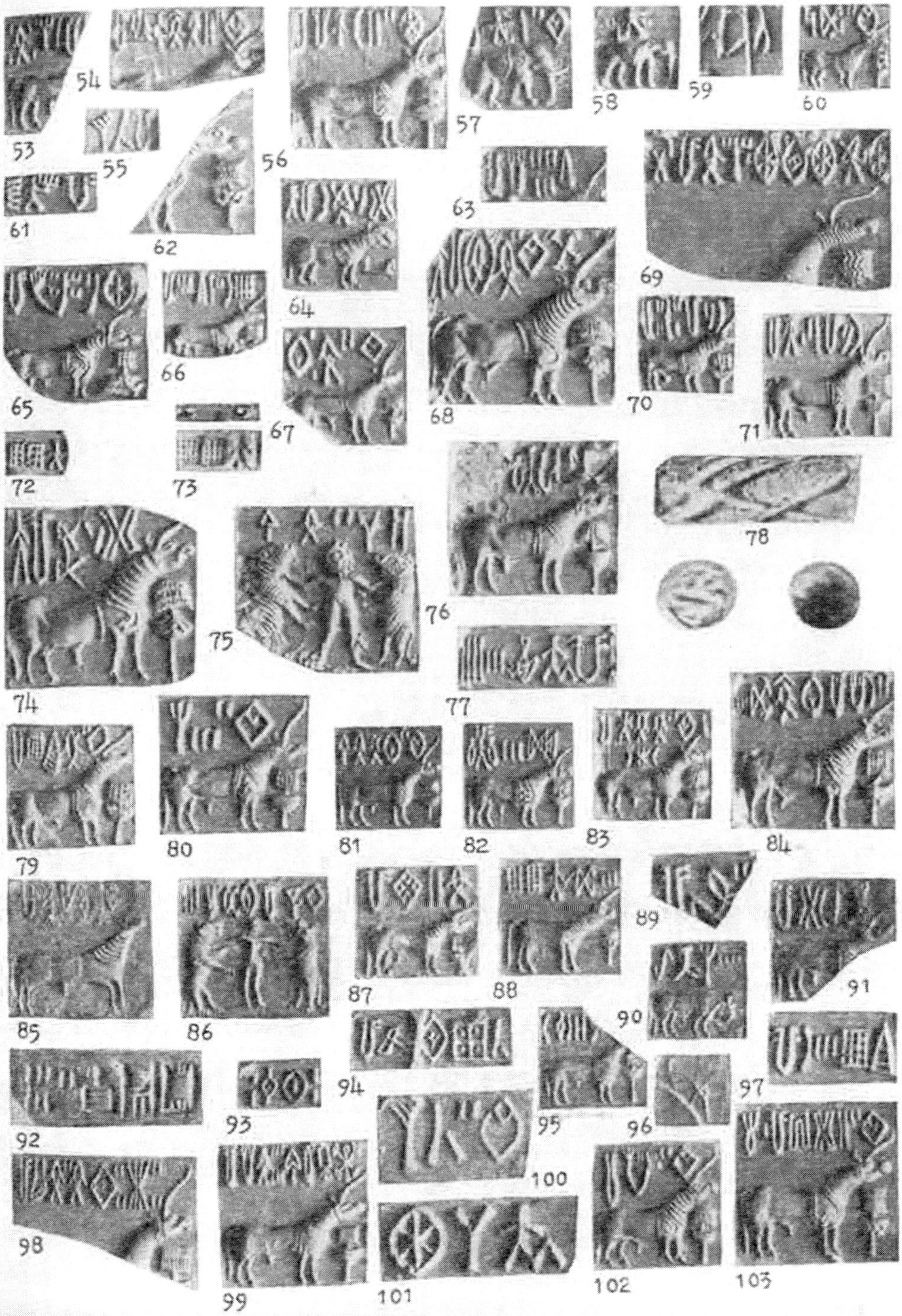

104
105
106
107
108
109
110
111
112
113
114
115
116
117
118
119
120
121
122
123
124
125
126
127
128
129
130
131
132
133
134
135
136
137
138
139
140
141
142
143
144
145
146
147
148
149
150
151
152
153
154
155

MOHENJO-DARO, DK. AREA, G SECTION.
IMPRESSIONS OF SEALS: UPPER LEVELS.

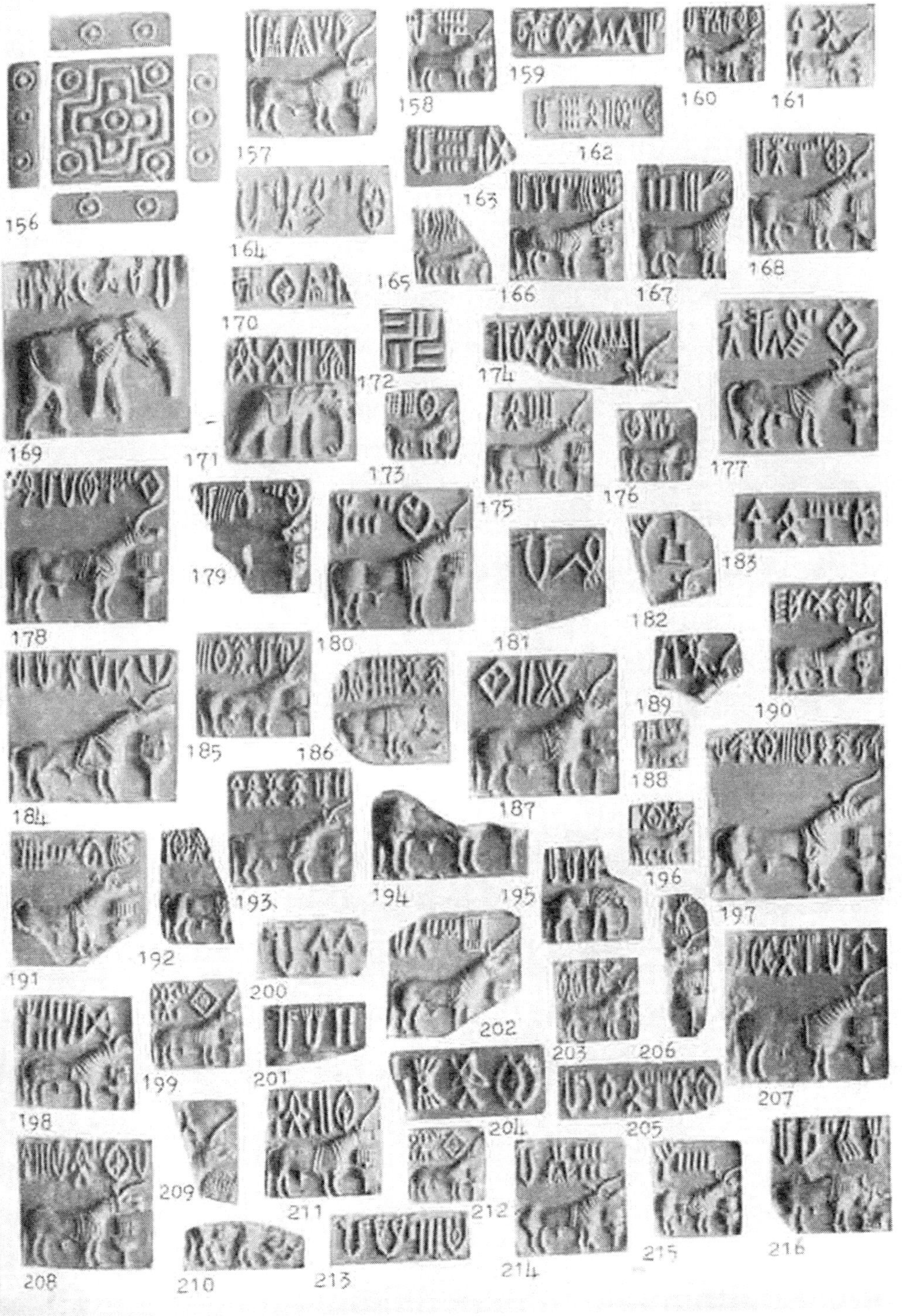

217
218
219
220
221
222
223
224
225
226
227
228
229
230
231
232
233
234
235
236
237
238
239
240
241
242
243
244
245
246
247
248
249
250
251
252
253
254
255
256
257
258
259
260
261
262
263
264
265
266
267
268
269
270
271

MOHENJO-DARO. DK. AREA, G. SECTION.
IMPRESSIONS OF SEALS, UPPER LEVELS.

IMPRESSIONS OF SEALS. UPPER LEVELS.

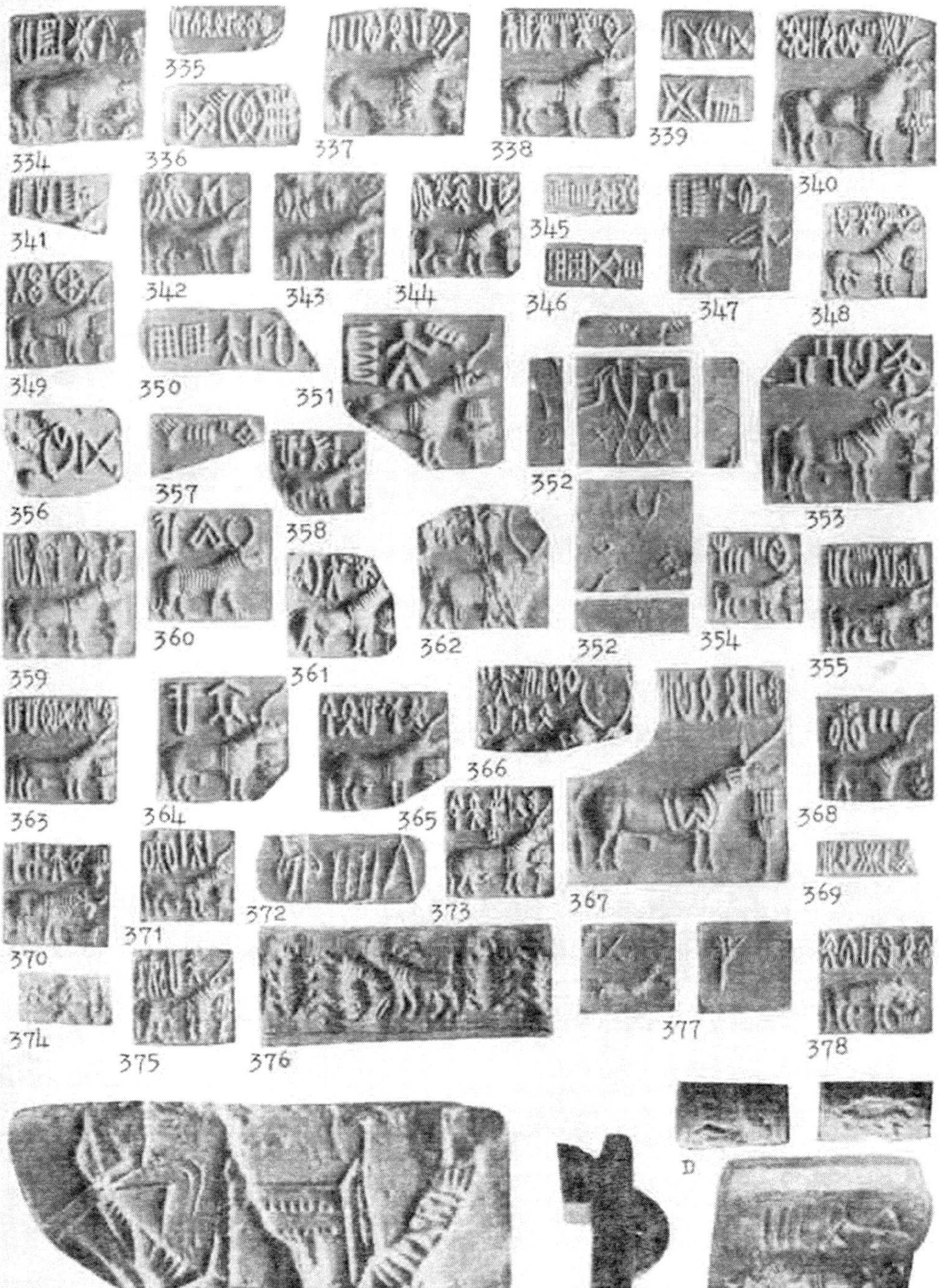

MOHENJO-DARO, DK. AREA, G SECTION.
SEALS AND AMULETS UPPER LEVELS.

MOHENJO-DARO, DK. AREA, G. SECTION. PLATE XCI.

CLAY AND FAIENCE AMULETS, COMBS, PINS AND BUTTONS, UPPER LEVELS.

MOHENJO-DARO, DK. AREA, G. SECTION.

CLAY AND FAIENCE AMULETS.

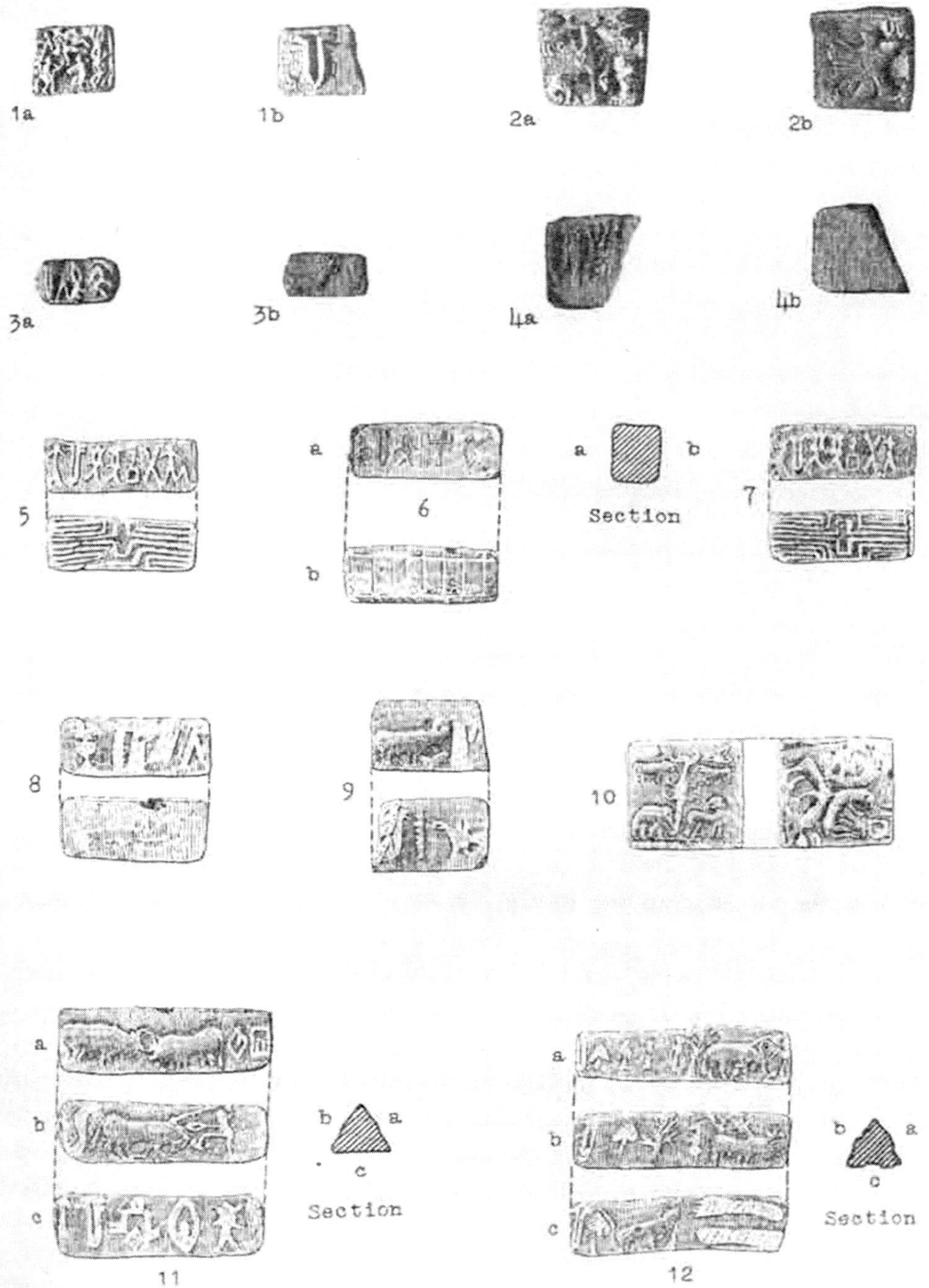

COPPER TABLETS.

MOHENJO-DARO, DK. AREA, G. SECTION.
IMPRESSIONS OF SEALS, LOWER LEVELS.

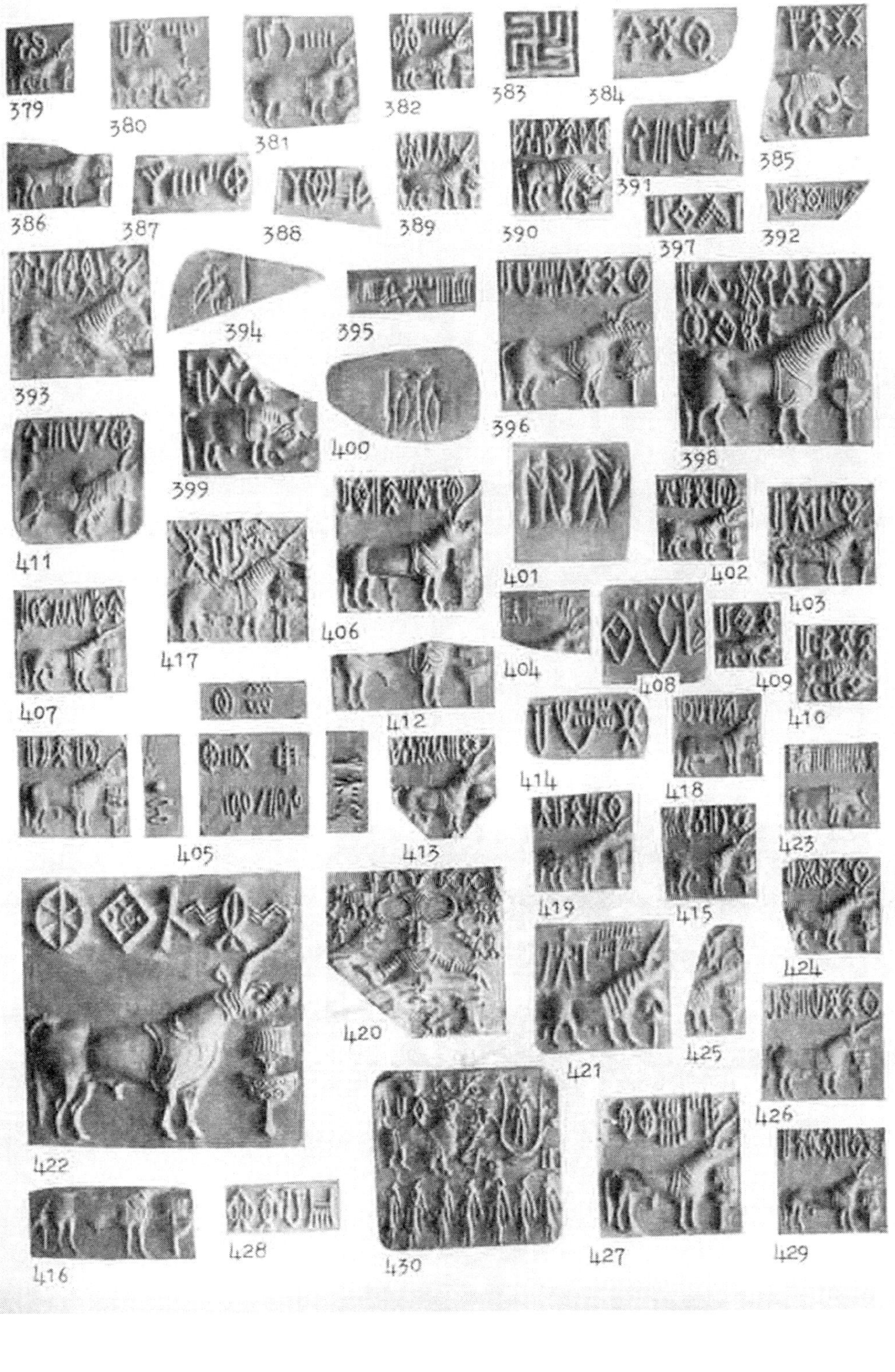

IMPRESSIONS OF SEALS. LOWER LEVELS.

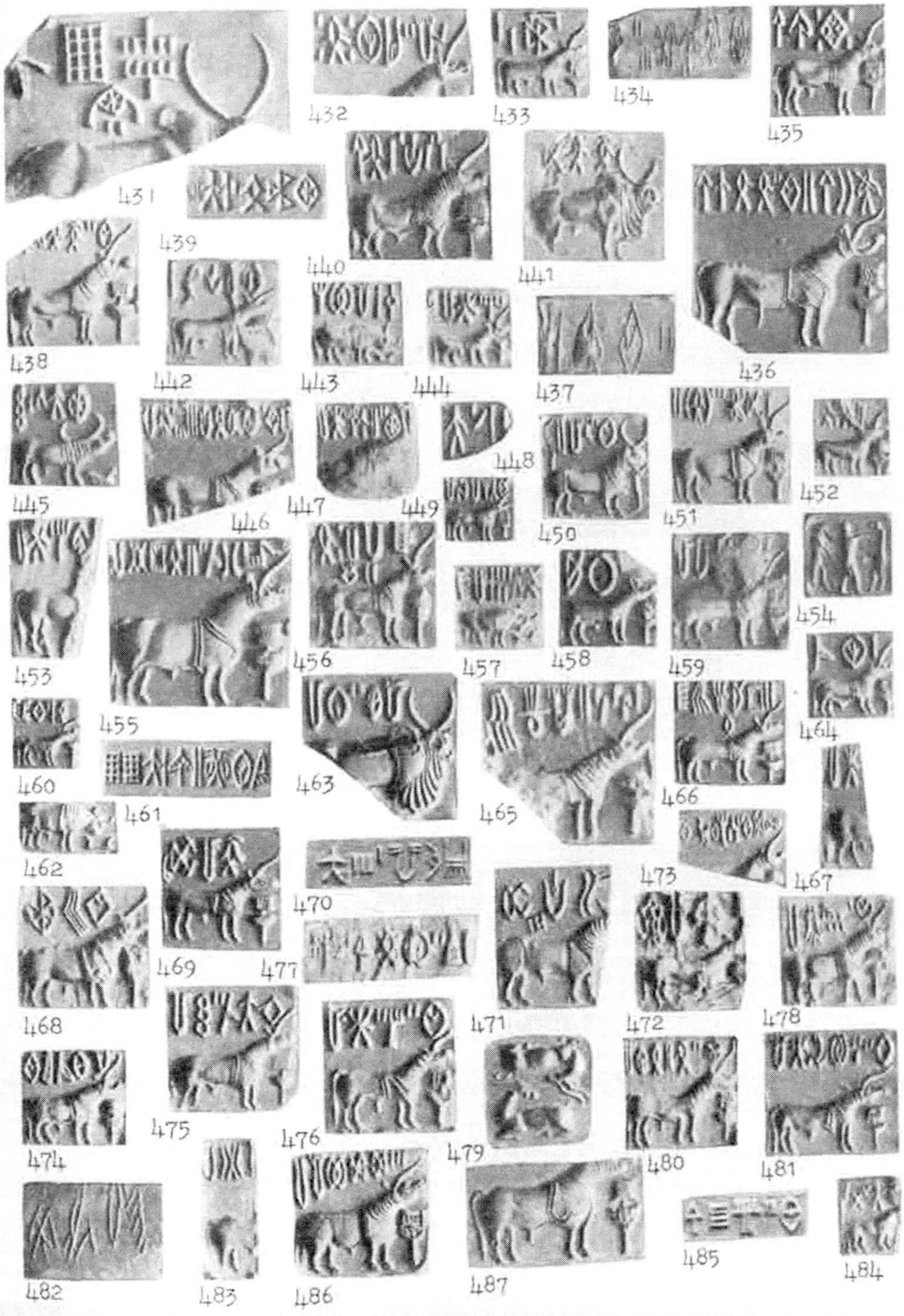

IMPRESSIONS OF SEALS. LOWER LEVELS.

IMPRESSIONS OF SEALS. LOWER LEVELS.

MOHENJO-DARO: DK. AREA, G. SECTION.

IMPRESSIONS OF SEALS. LOWER LEVELS.

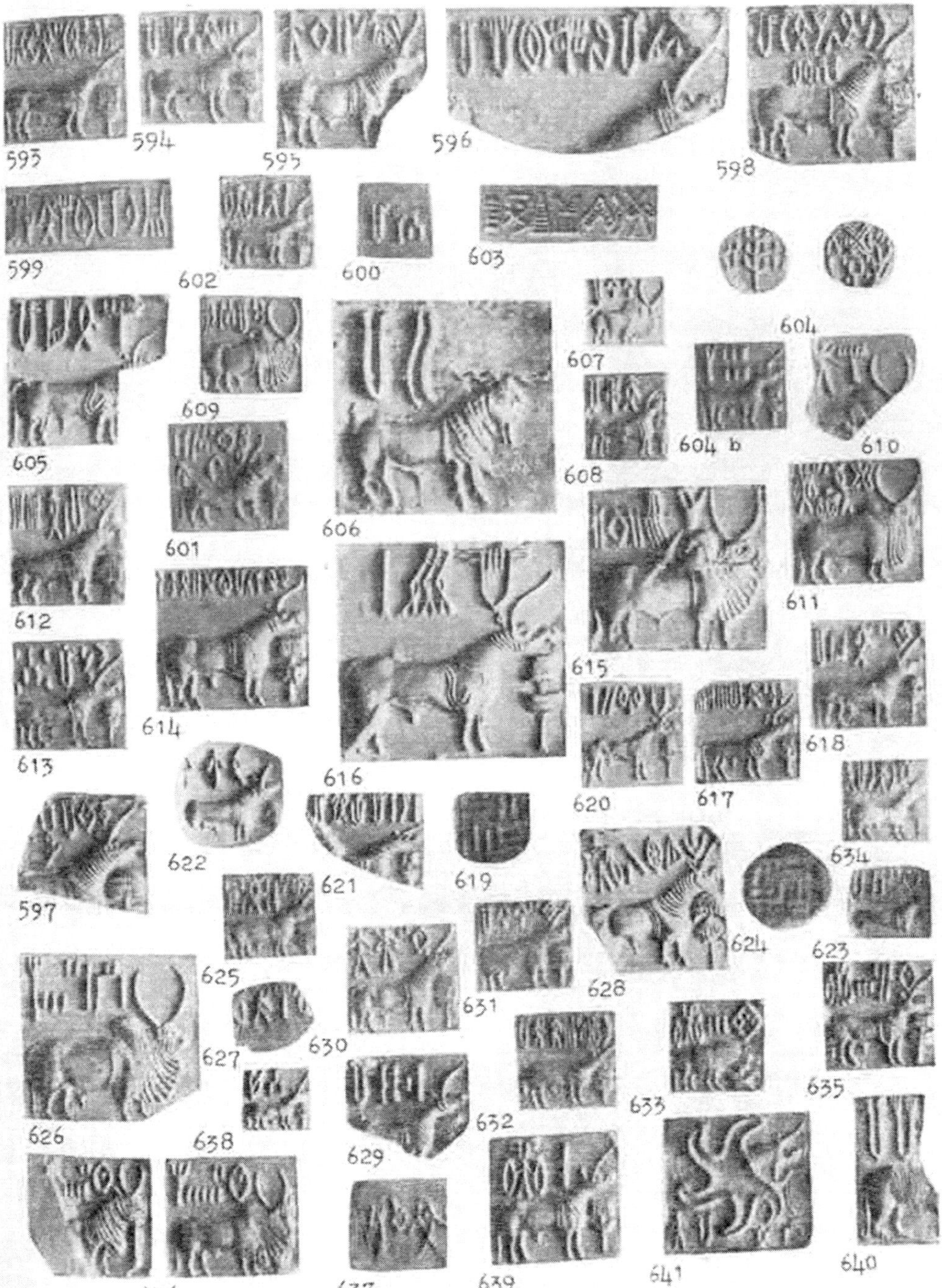

MOHENJO-DARO, DK. AREA, G. SECTION.
IMPRESSIONS OF SEALS. LOWER LEVELS.

MOHENJO-DARO, DK. AREA, G. SECTION. PLATE C.

SEAL IMPRESSIONS, HAIR-PINS, COMB, BUTTONS, ETC. LOWER LEVELS.

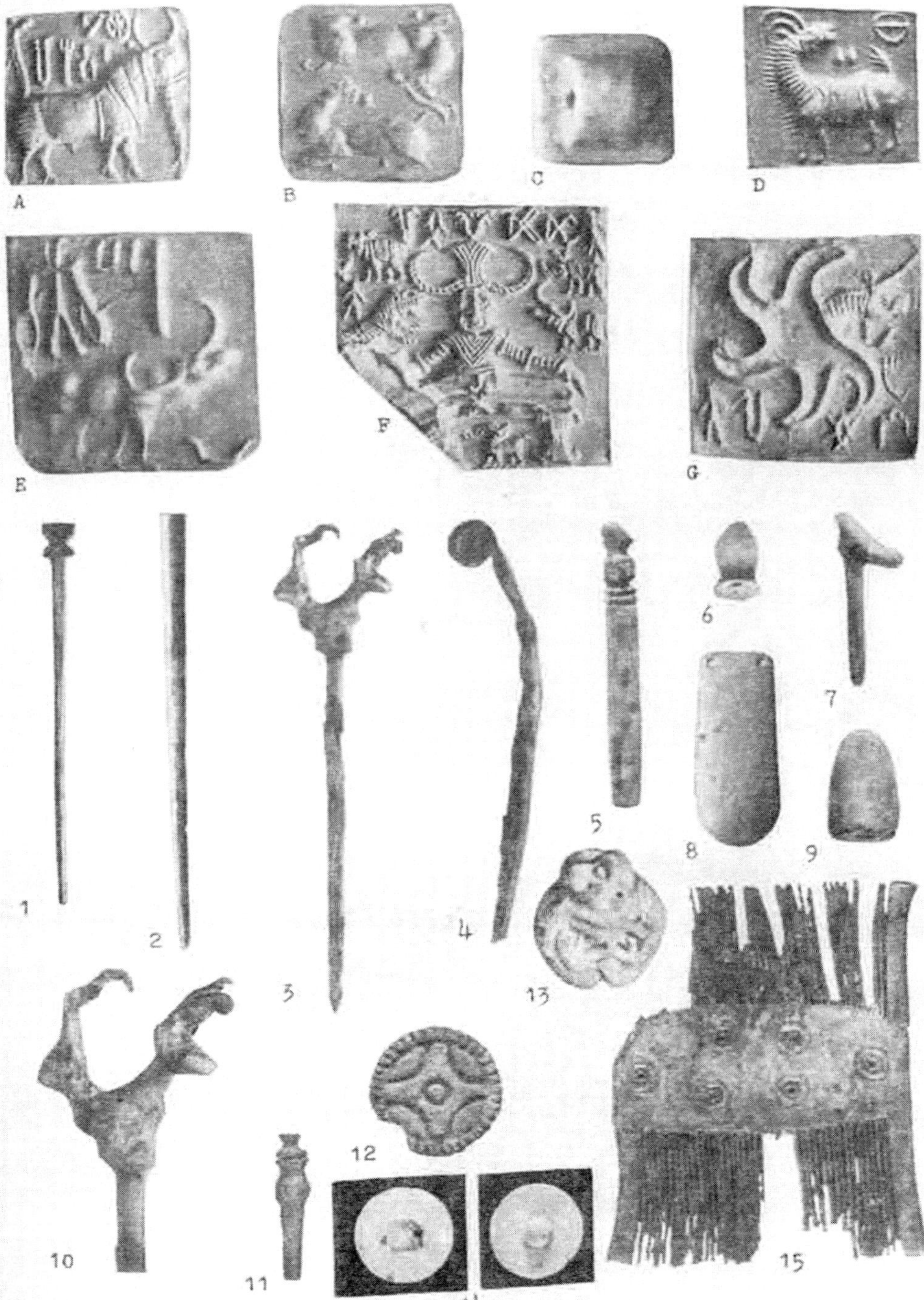

MOHENJO-DARO, DK AREA, G. SECTION.

CLAY AND FAIENCE AMULETS.

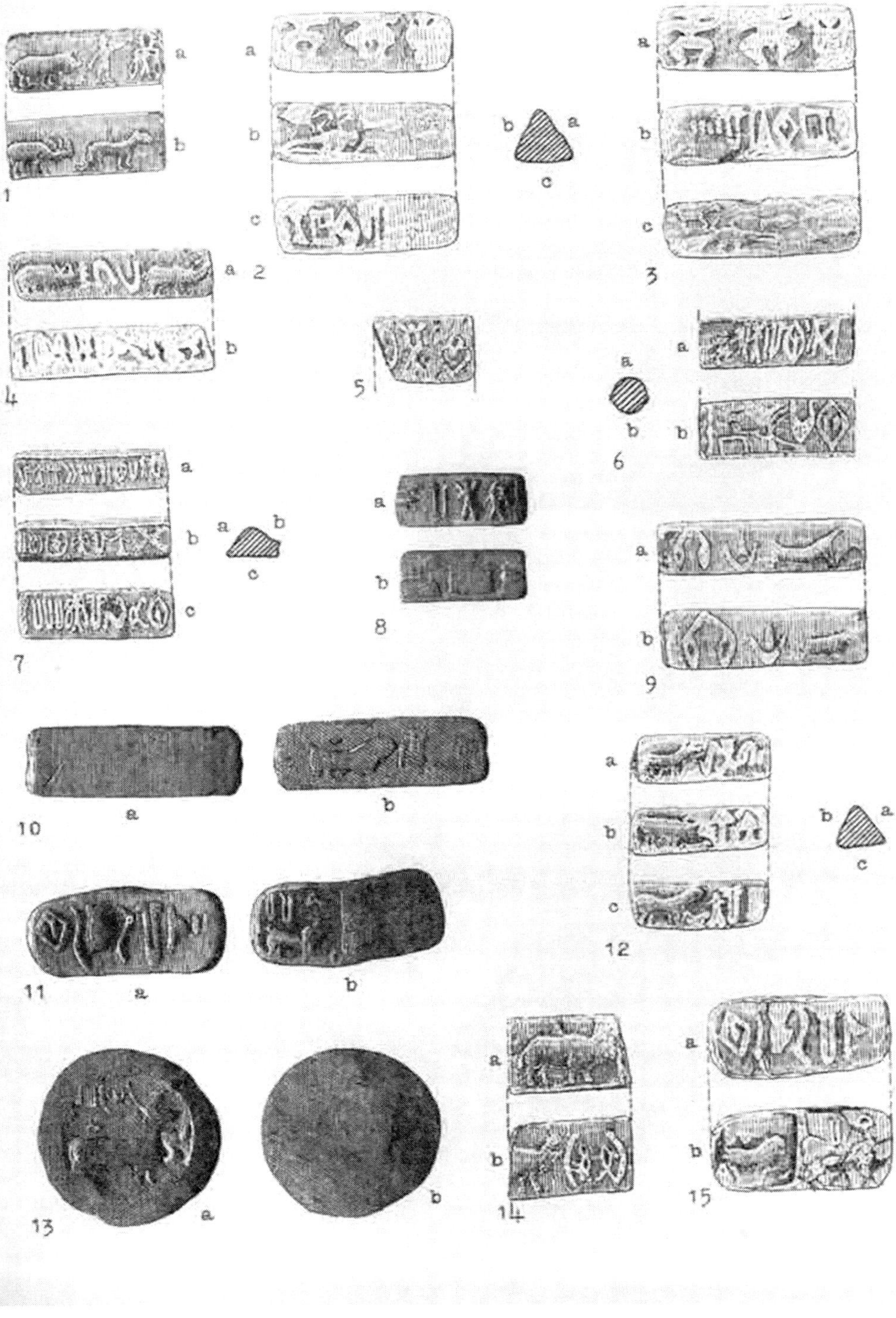

MOHENJO-DARO, DK. AREA, G SECTION.

CLAY AND FAIENCE AMULETS.

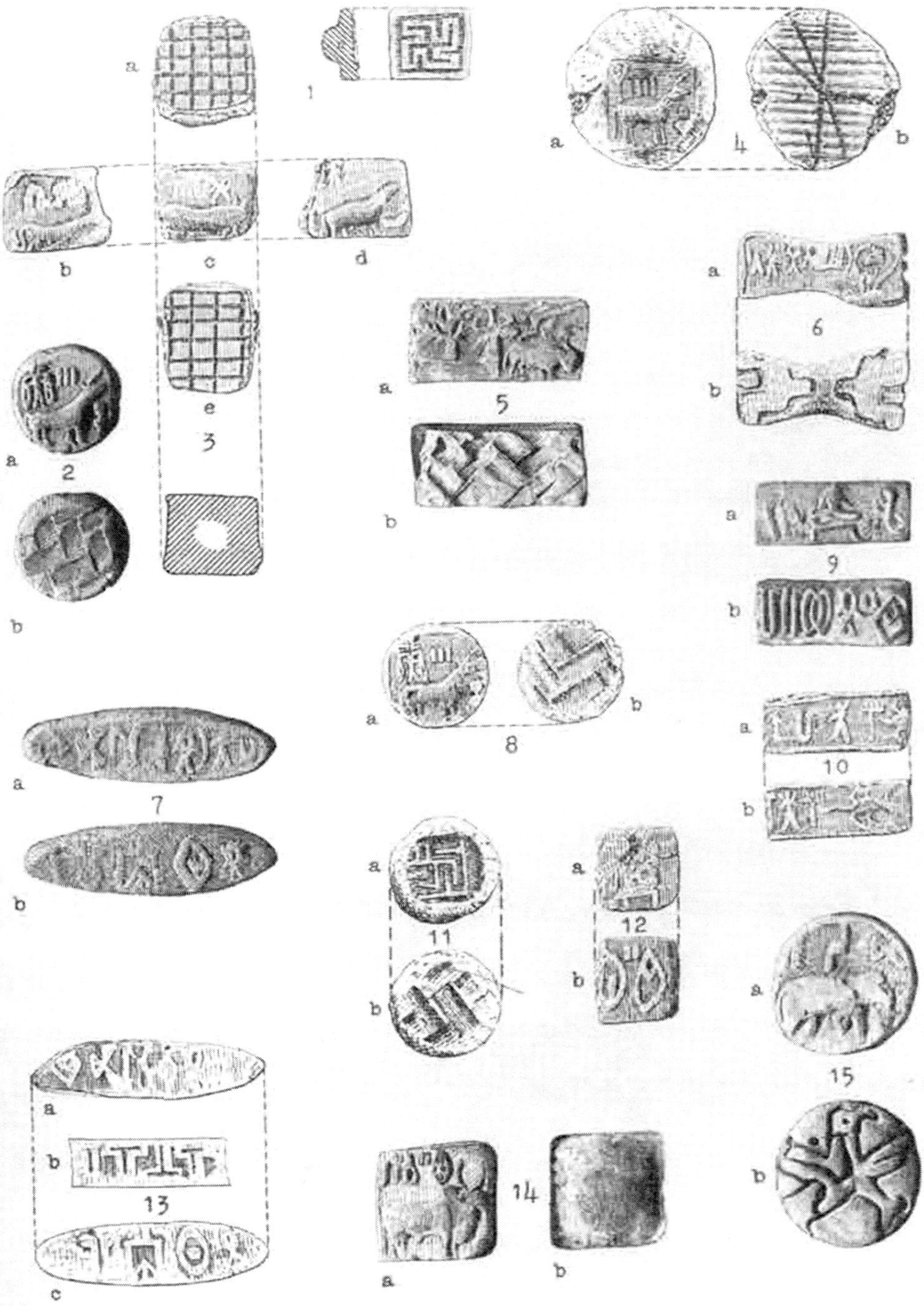

MOHENJO-DARO, DK. AREA. G. SECTION.

COPPER TABLETS AND CLAY AMULETS, LOWER LEVELS.

MOHENJO-DARO, DK. AREA, G. SECTION.
STONE AND POTTERY OBJECTS.

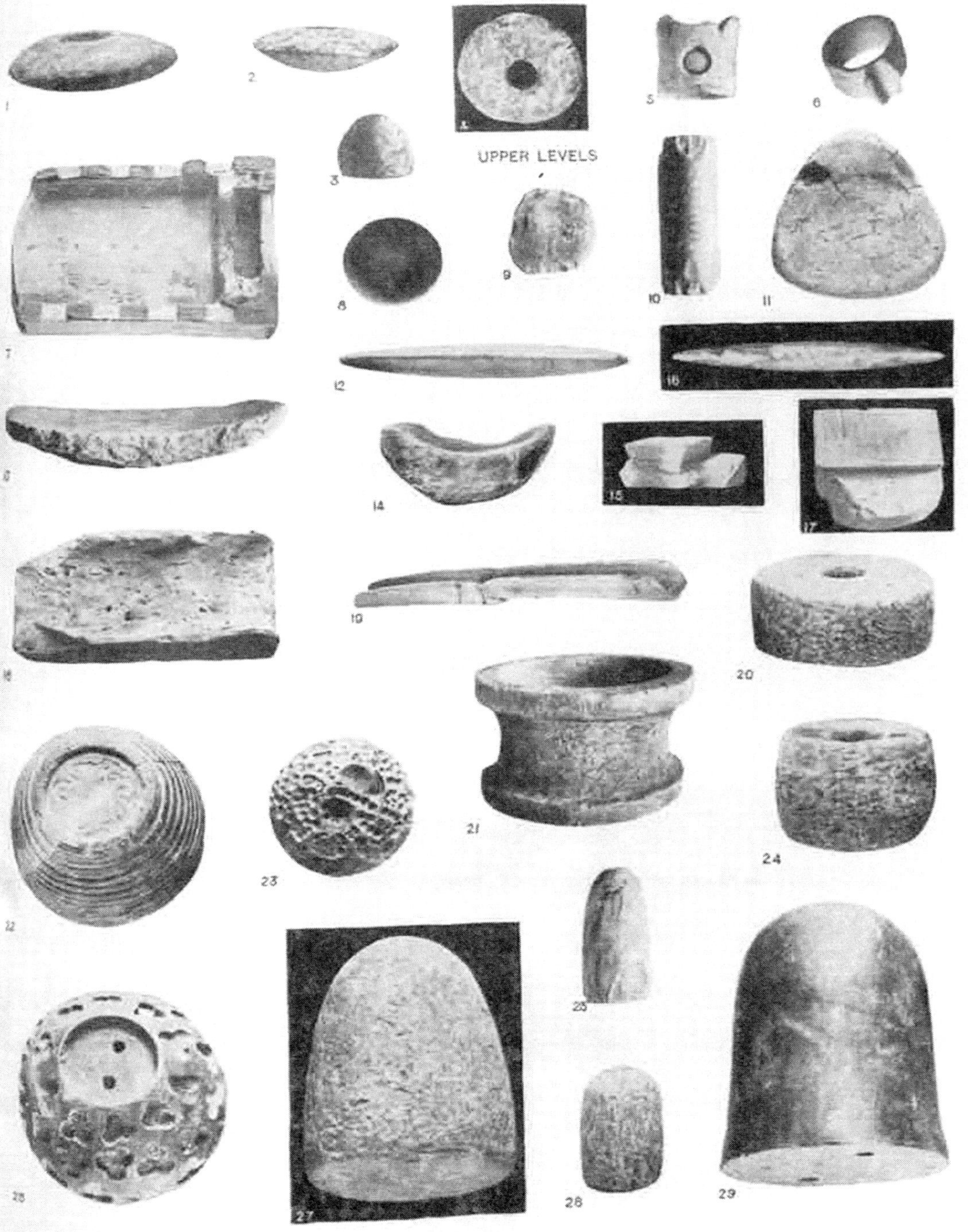

MOHENJO-DARO, DK. AREA, G. SECTION.
MISCELLANEOUS OBJECTS FROM UPPER LEVELS.

MISCELLANEOUS OBJECTS.

MOHENJO-DARO, DK AREA, G. SECTION.
MISCELLANEOUS OBJECTS.

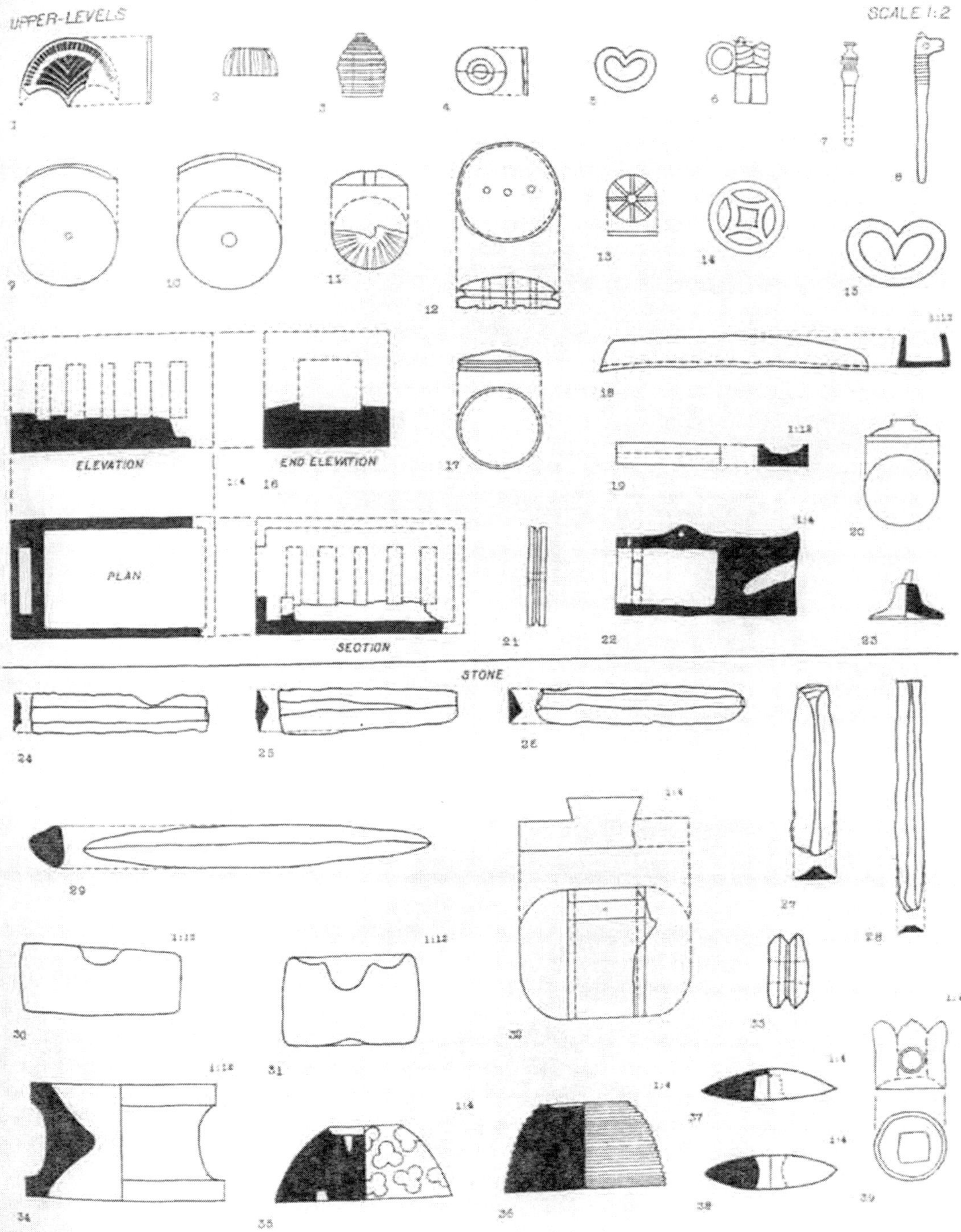

MOHENJO-DARO, DK. AREA. G. SECTION.
MISCELLANEOUS OBJECTS.

MOHENJO-DARO, DK. AREA, G. SECTION.
MISCELLANEOUS OBJECTS.

PLATE CX.

MOHENJO-DARO, DK. AREA, G. SECTION.
DRAWINGS OF MISCELLANEOUS OBJECTS.

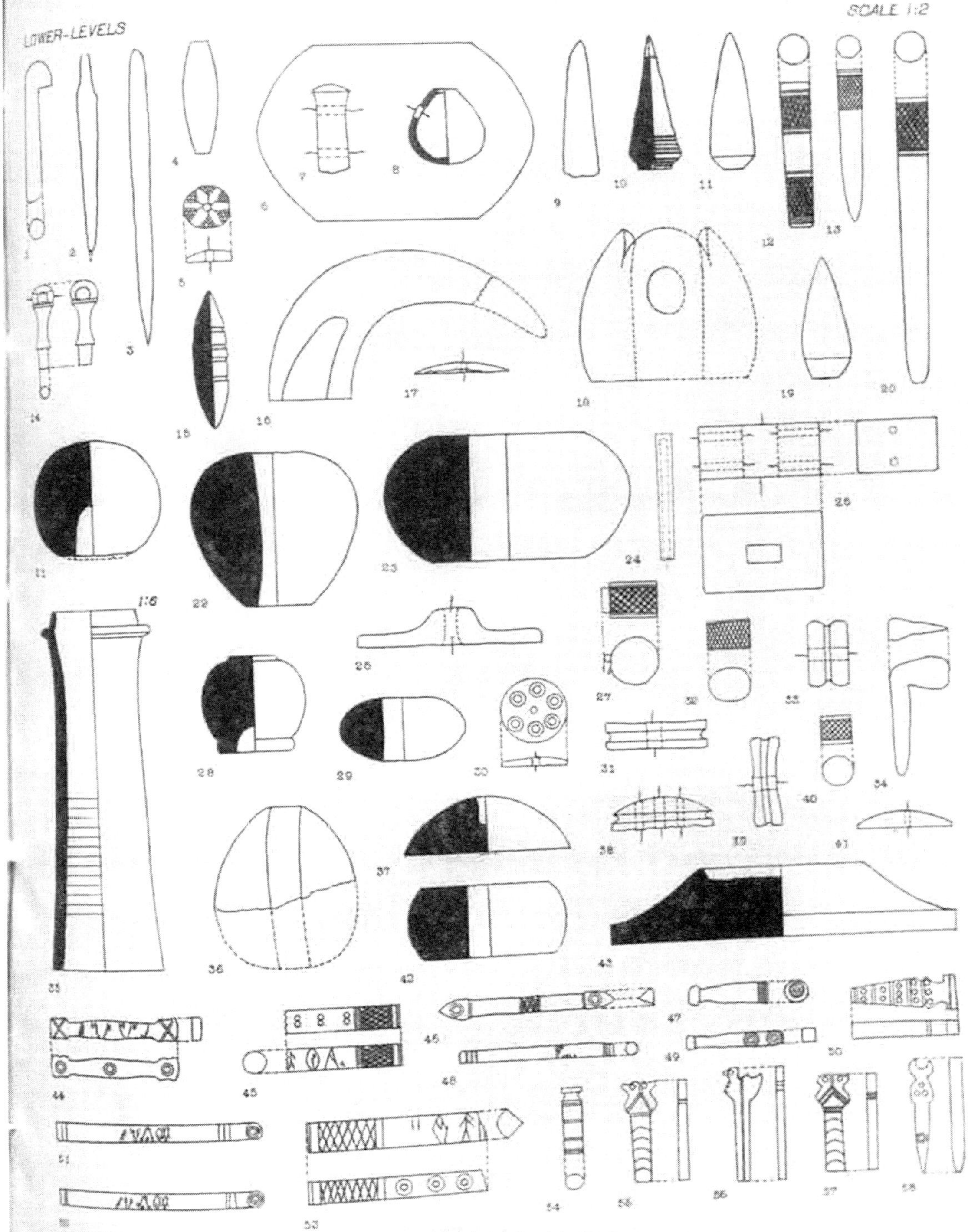

MOHENJO-DARO, DK. AREA, G. SECTION,
MISCELLANEOUS OBJECTS.

MOHENJO-DARO: DK. AREA, G. SECTION.

POTTERY OBJECTS, EARLY LEVELS.

MOHENJO-DARO, DK. AREA, G. SECTION. NORTHERN PORTION.

COPPER AND BRONZE TOOLS AND IMPLEMENTS:

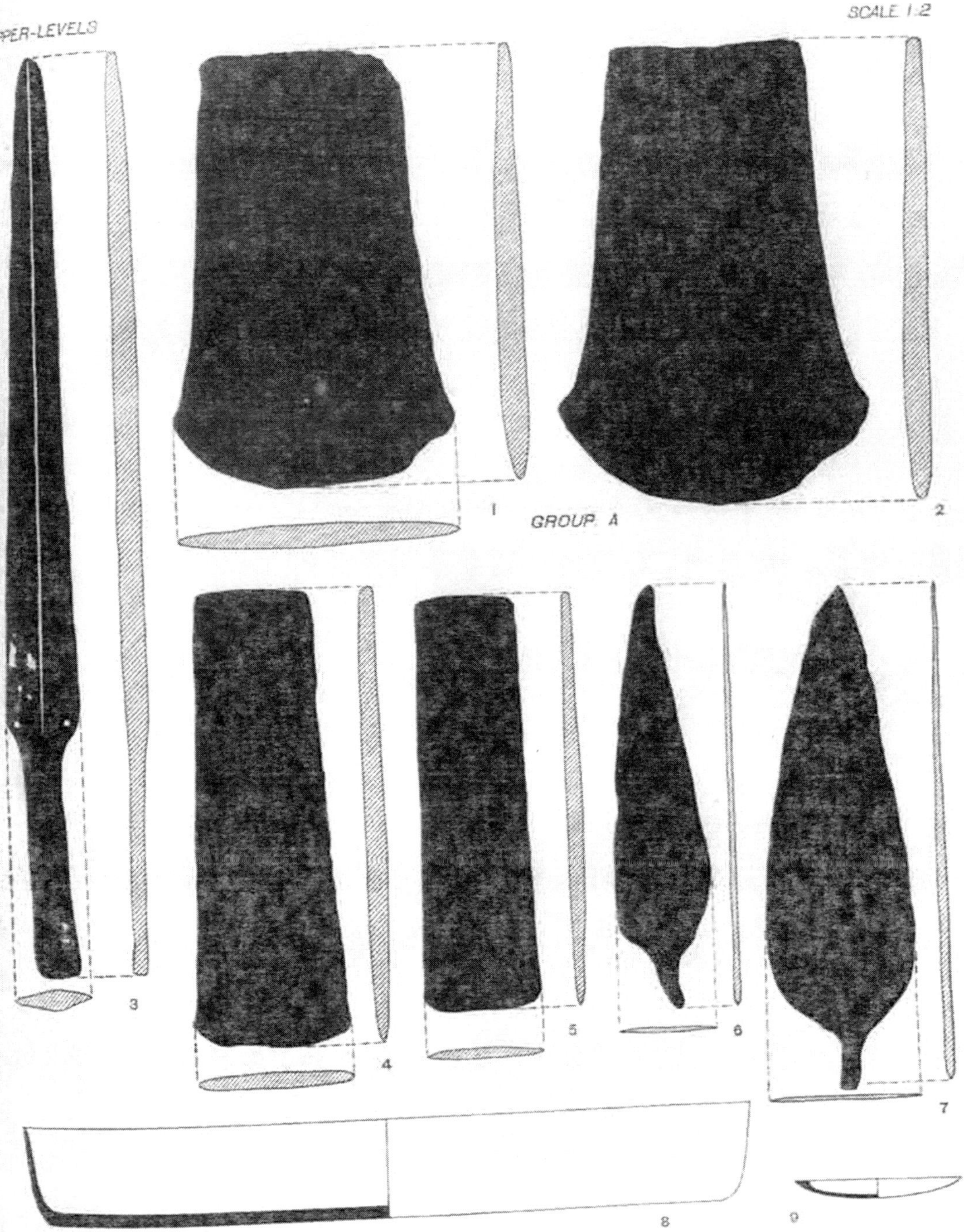

MOHENJO-DARO, DK. AREA, G. SECTION.

COPPER AND BRONZE TOOLS, IMPLEMENTS AND UTENSILS.

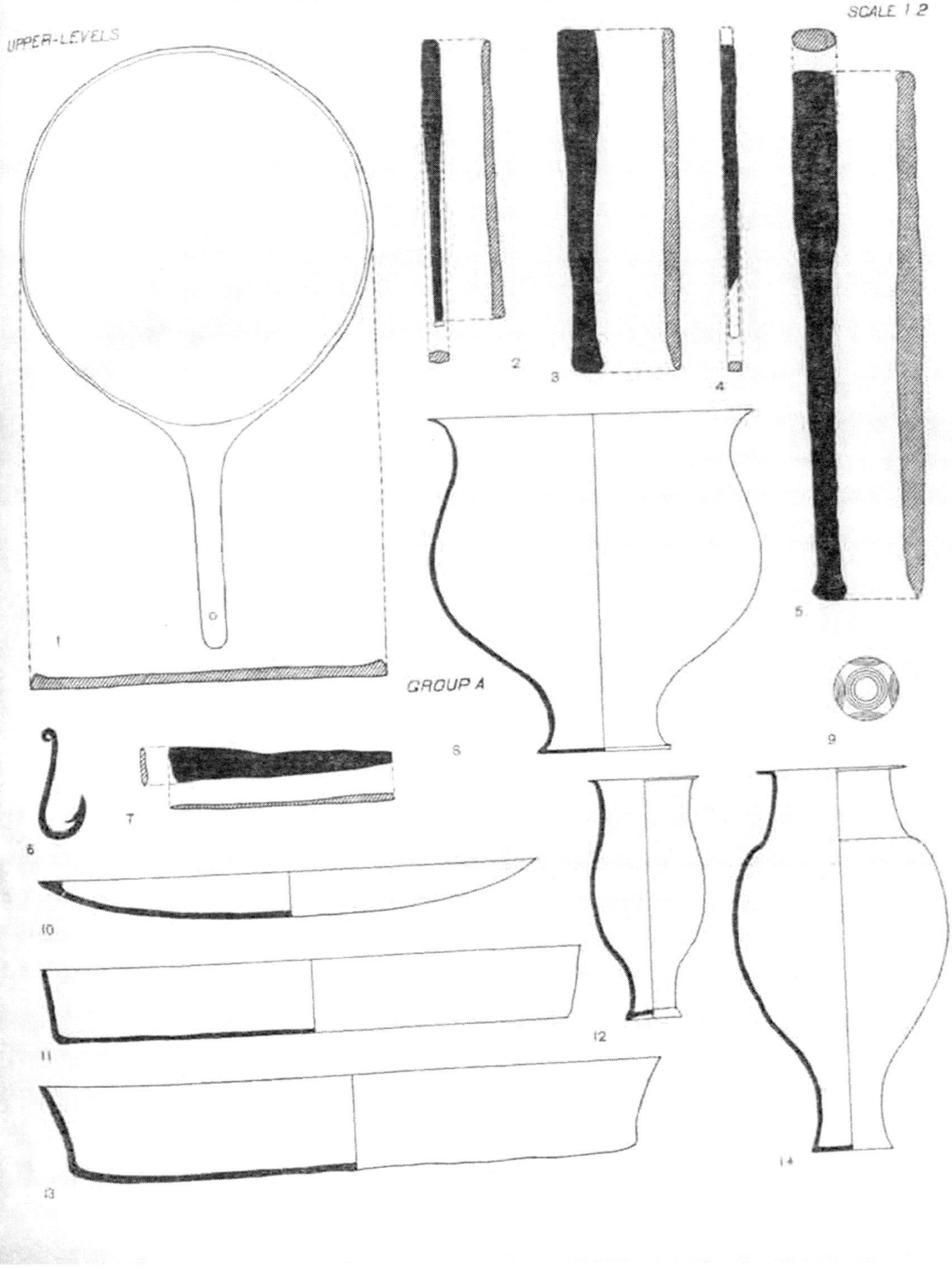

MOHENJO-DARO, DK AREA, G. SECTION. NORTHERN PORTION. PLATE CXV.

COPPER AND BRONZE UTENSILS, TOOLS AND STONE WEIGHTS.

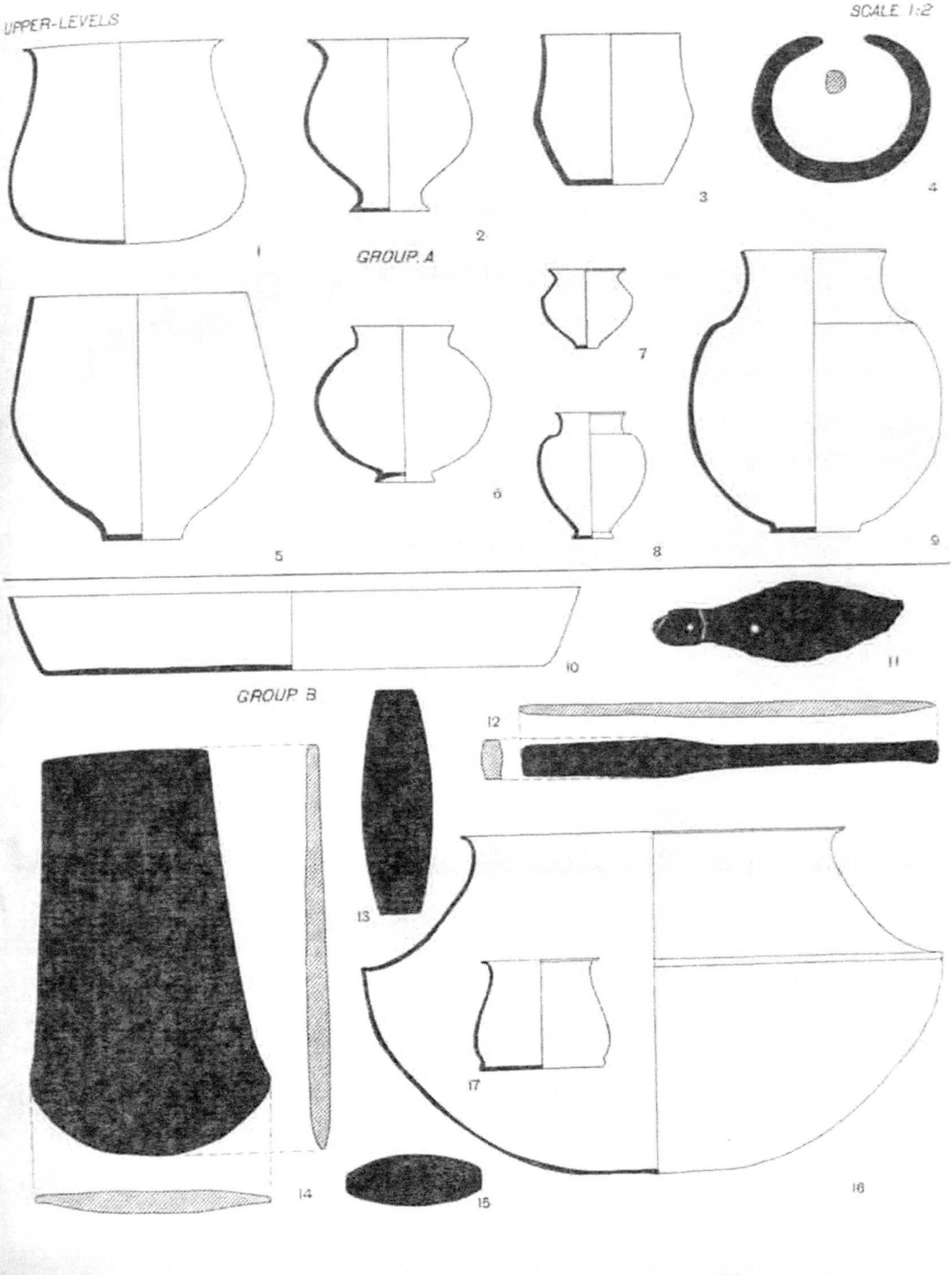

MOHENJO-DARO, DK. AREA, G. SECTION. NORTHERN PORTION.
COPPER AND BRONZE TOOLS AND UTENSILS AND STONE VESSEL.

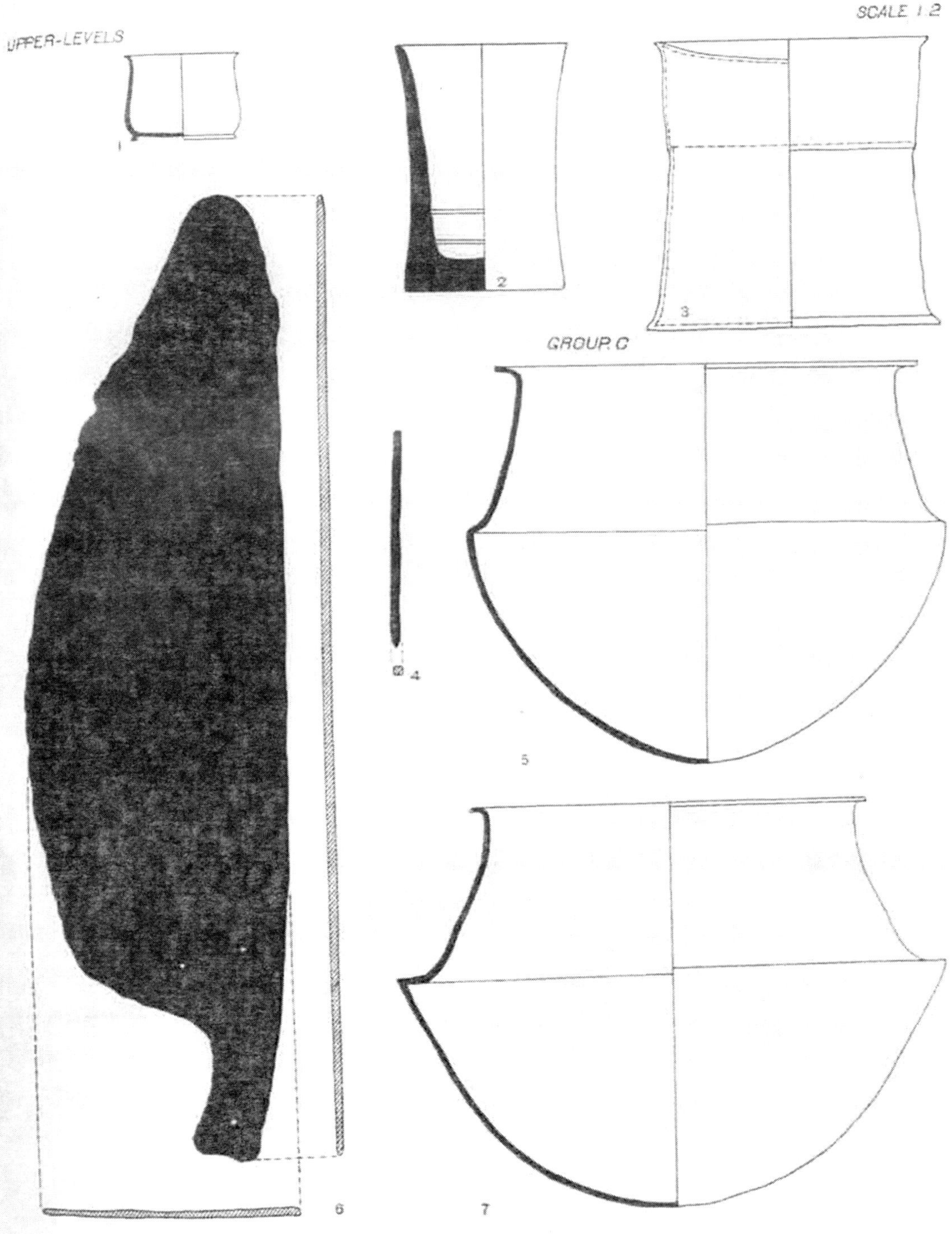

COPPER AND BRONZE TOOLS.

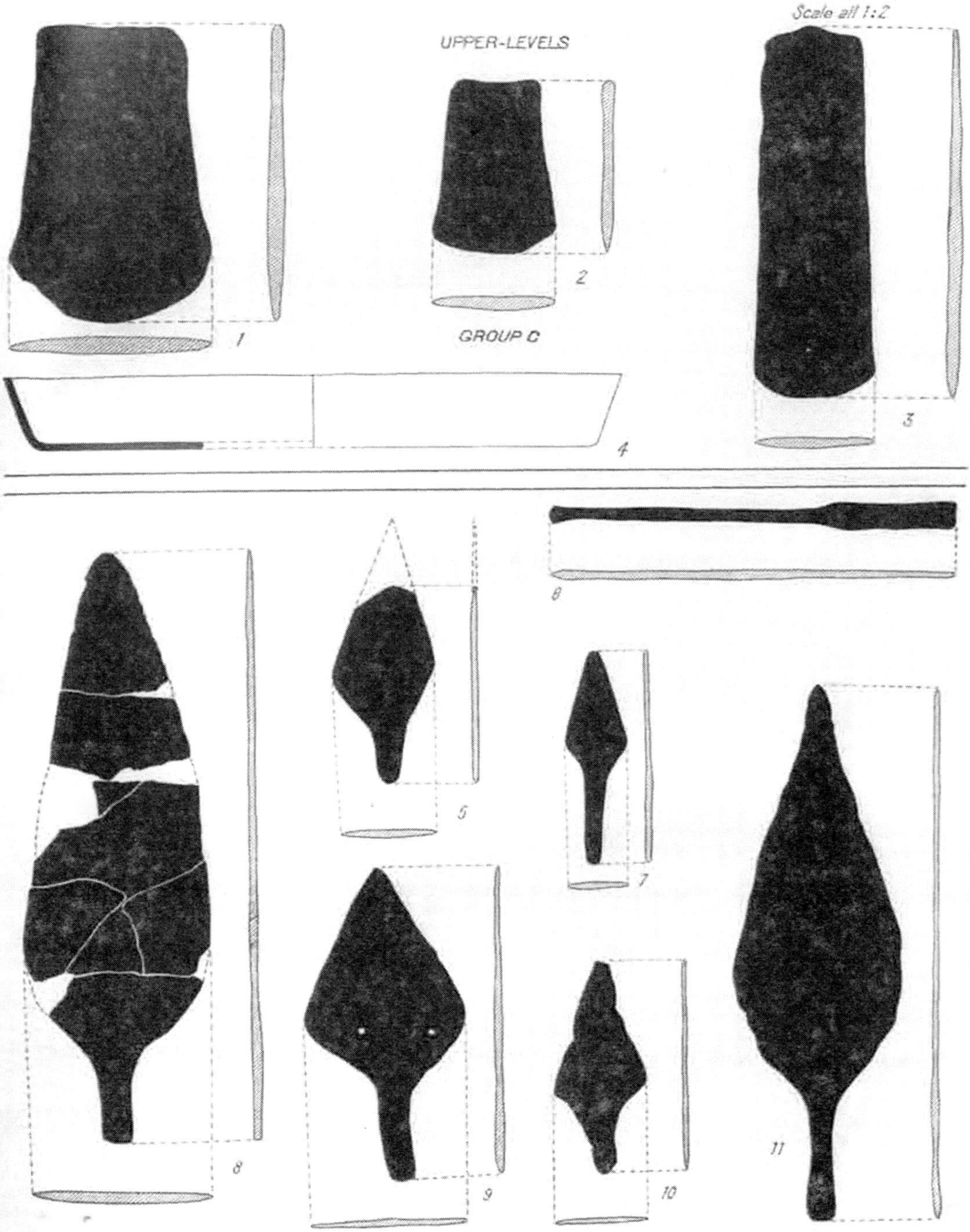

MOHENJO-DARO, DK. AREA, G SECTION,
COPPER AND BRONZE OBJECTS, UPPER LEVELS

COPPER AND BRONZE TOOLS AND IMPLEMENTS.

UPPER-LEVELS

SCALE 1:2

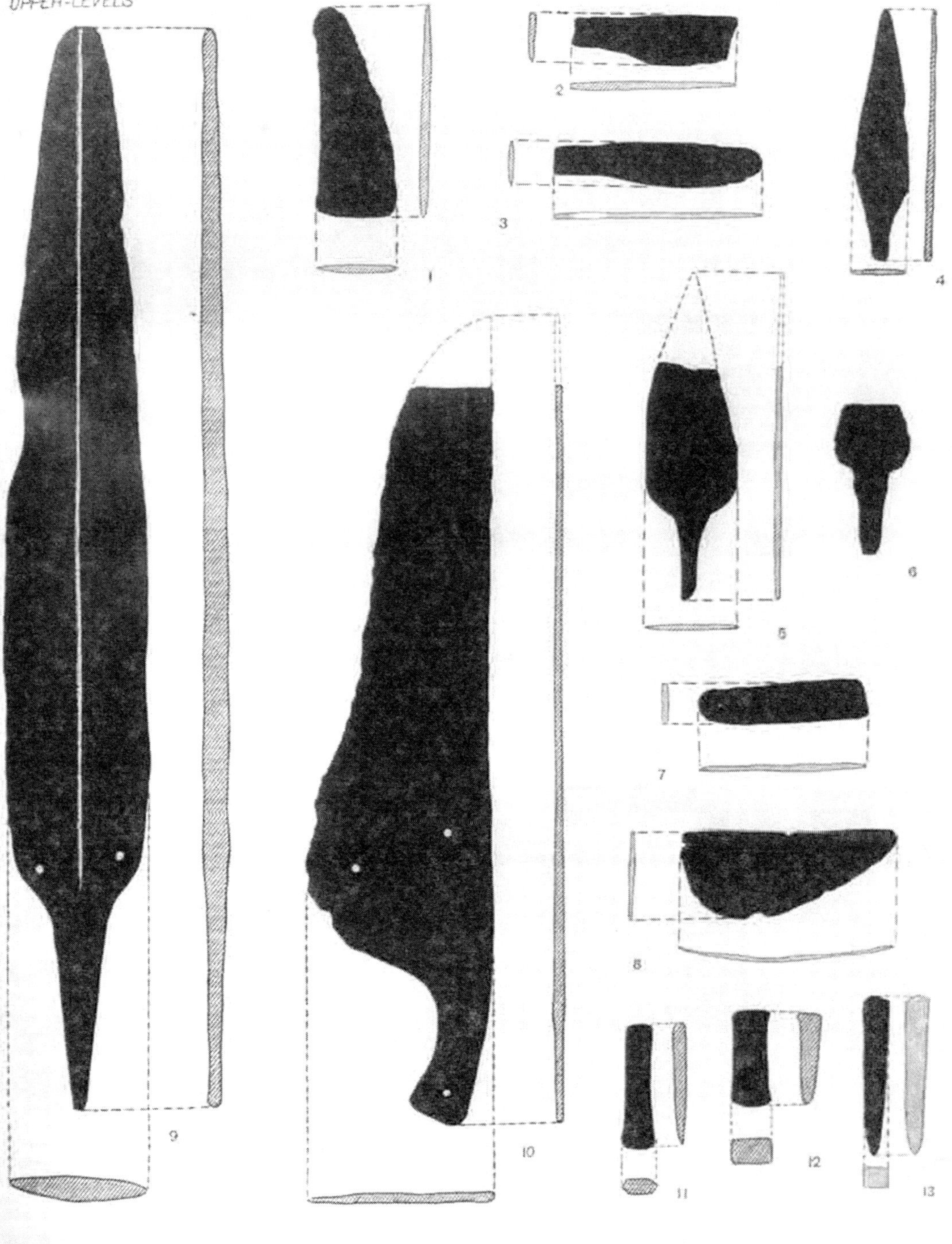

MOHENJO-DARO: DK. AREA, G. SECTION.
COPPER AND BRONZE IMPLEMENTS. UPPER LEVELS.

MOHENJO-DARO. DK. AREA, G. SECTION.
COPPER AND BRONZE TOOLS, IMPLEMENTS AND ORNAMENTS.

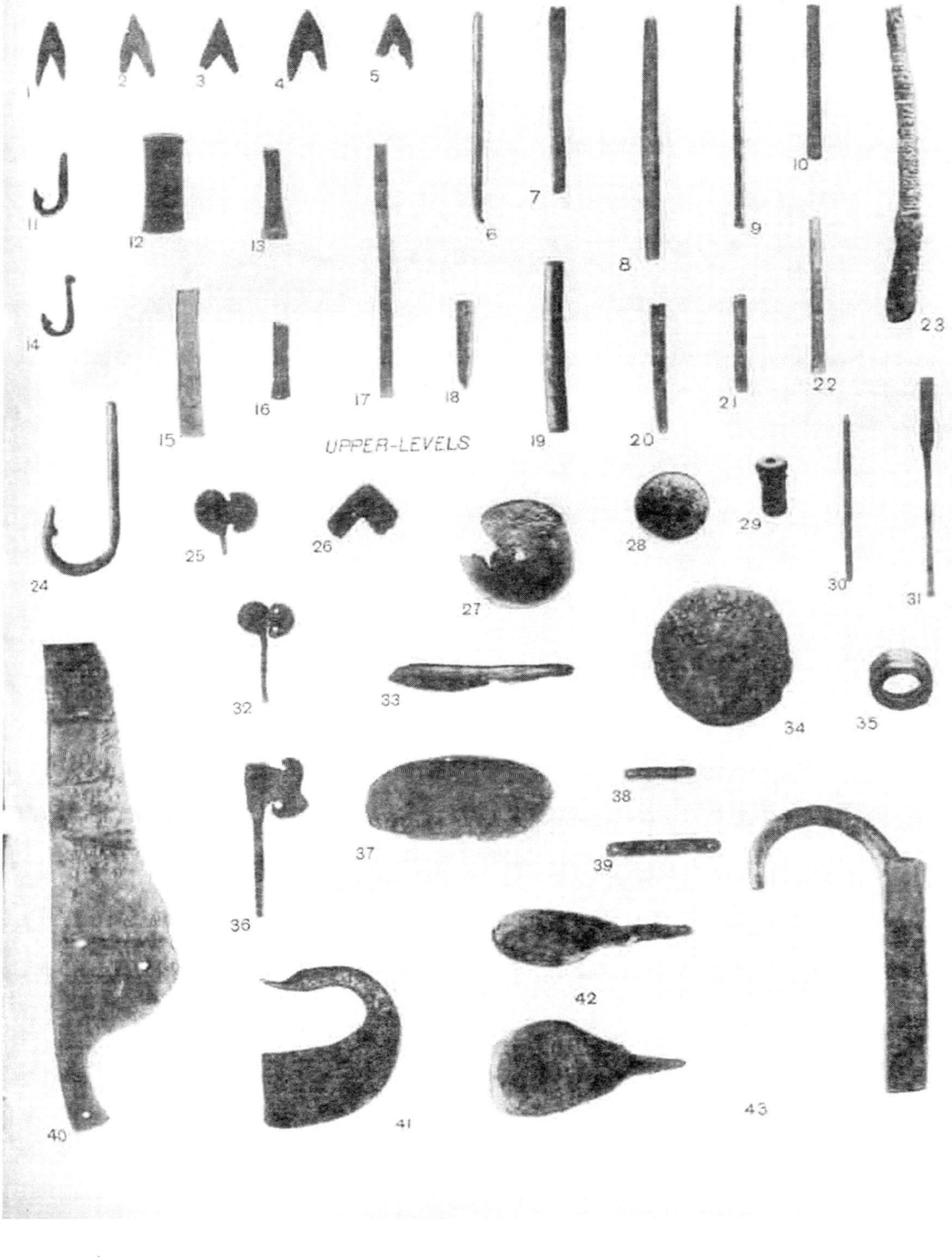

COPPER AND BRONZE TOOLS AND IMPLEMENTS.

COPPER AND BRONZE KNIVES AND DAGGERS.

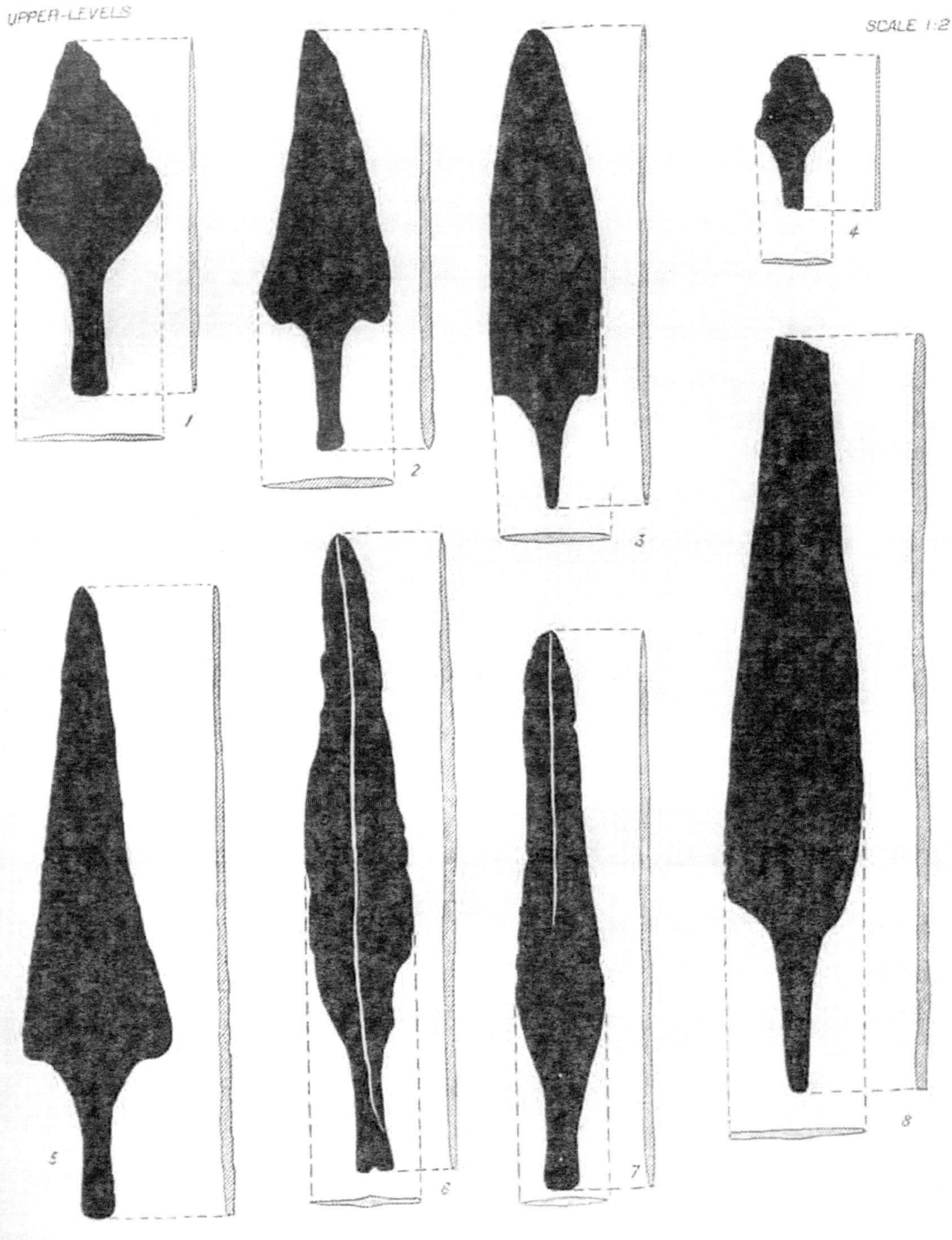

MOHENJO-DARO, DK AREA, G. SECTION

COPPER AND BRONZE OBJECTS.

SCALE 1:2

UPPER-LEVELS

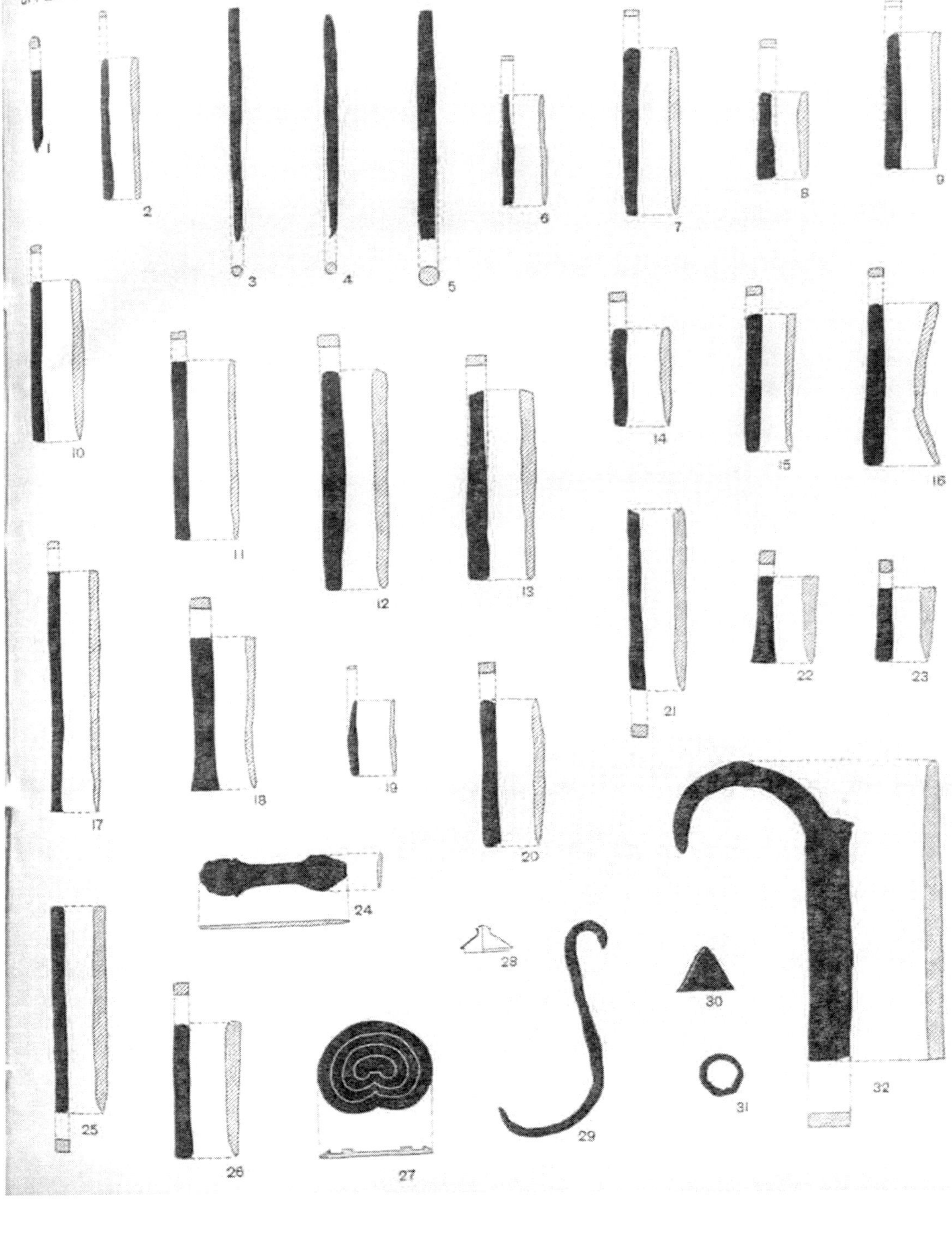

MISCELLANEOUS OBJECTS.

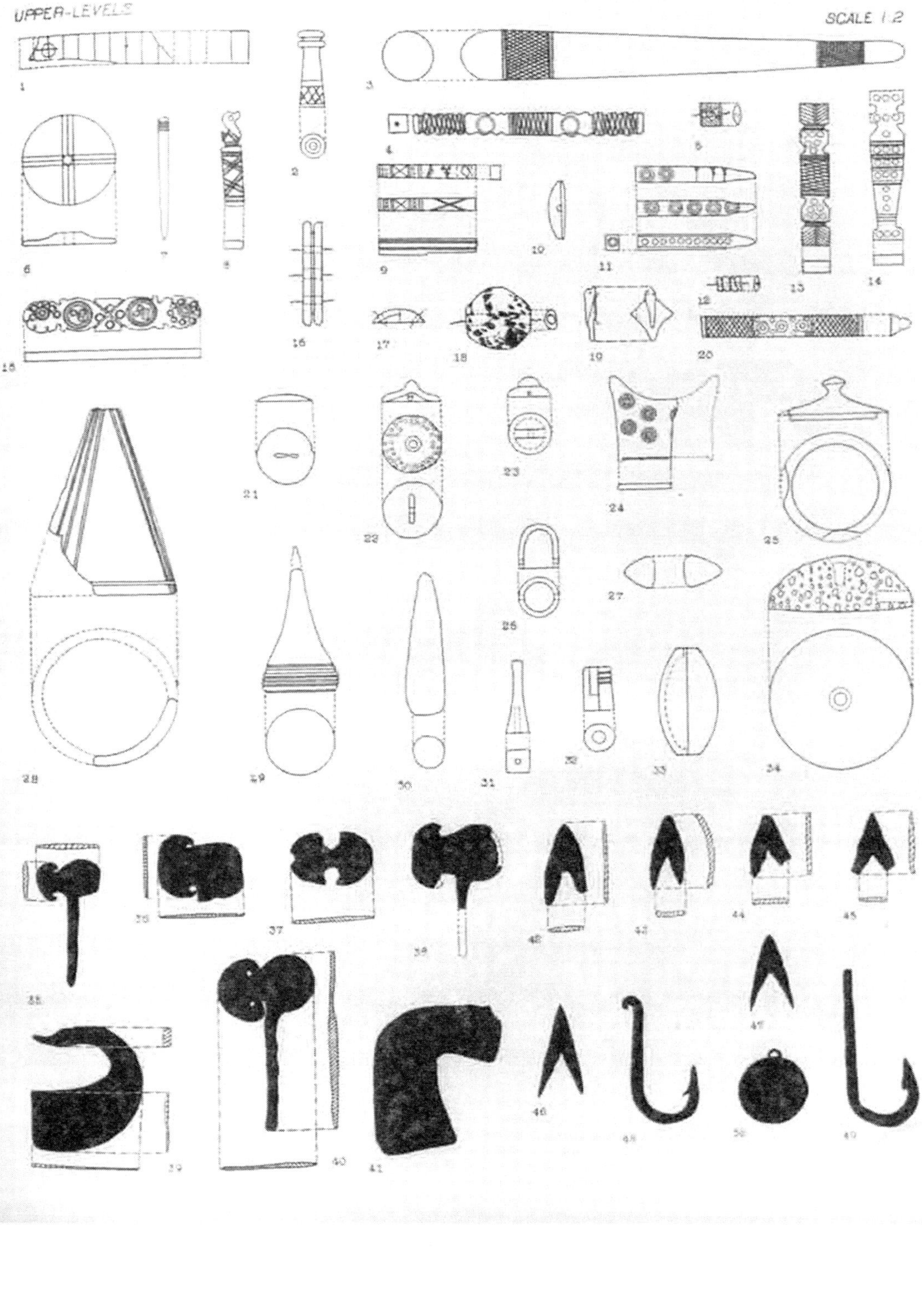

MOHENJO-DARO. DK. AREA, G. SECTION.
COPPER AND BRONZE IMPLEMENTS AND TOOLS.

LOWER-LEVELS

SCALE 1:2

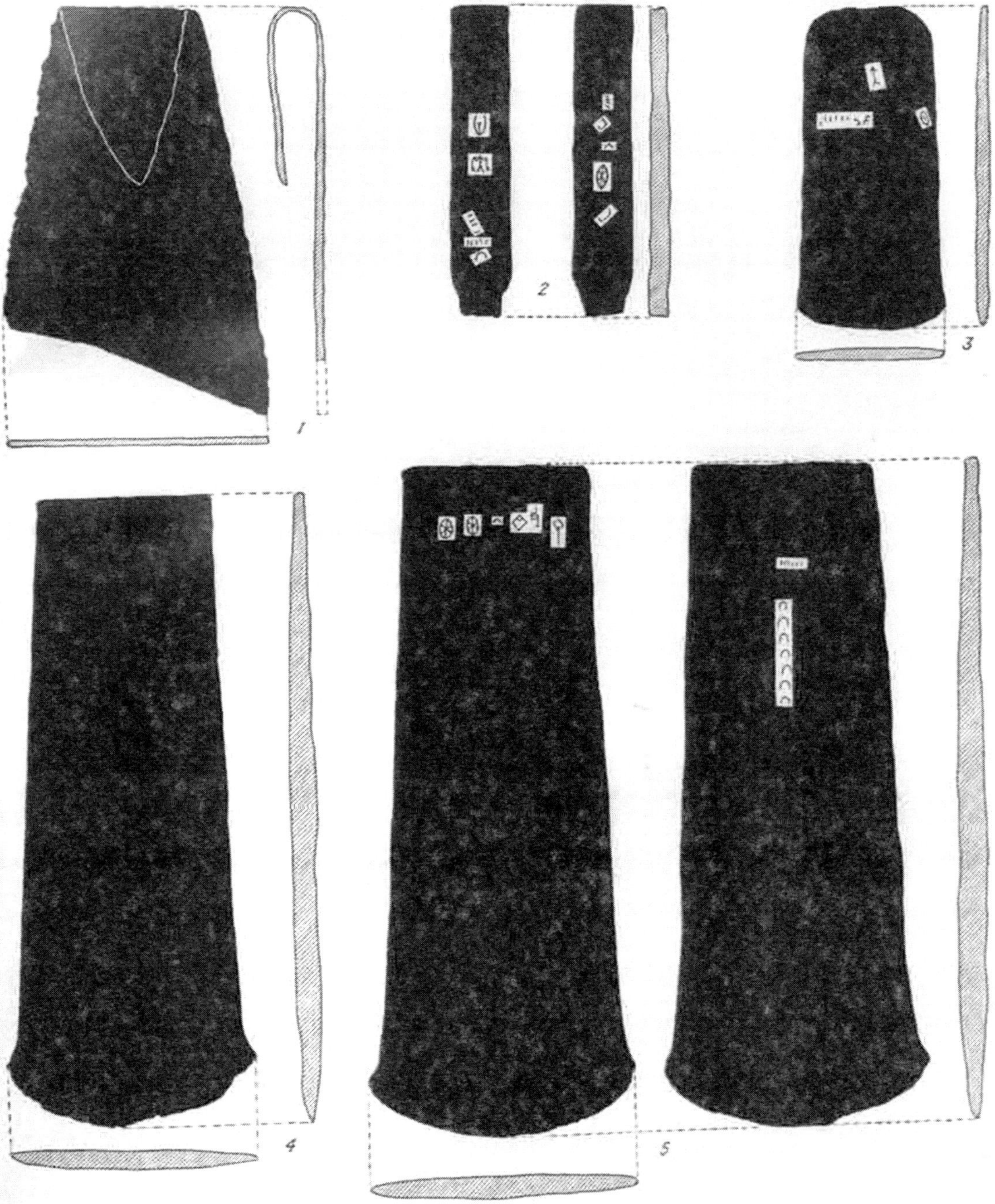

MOHENJO-DARO, DK. AREA, G. SECTION.

COPPER AND BRONZE OBJECTS.

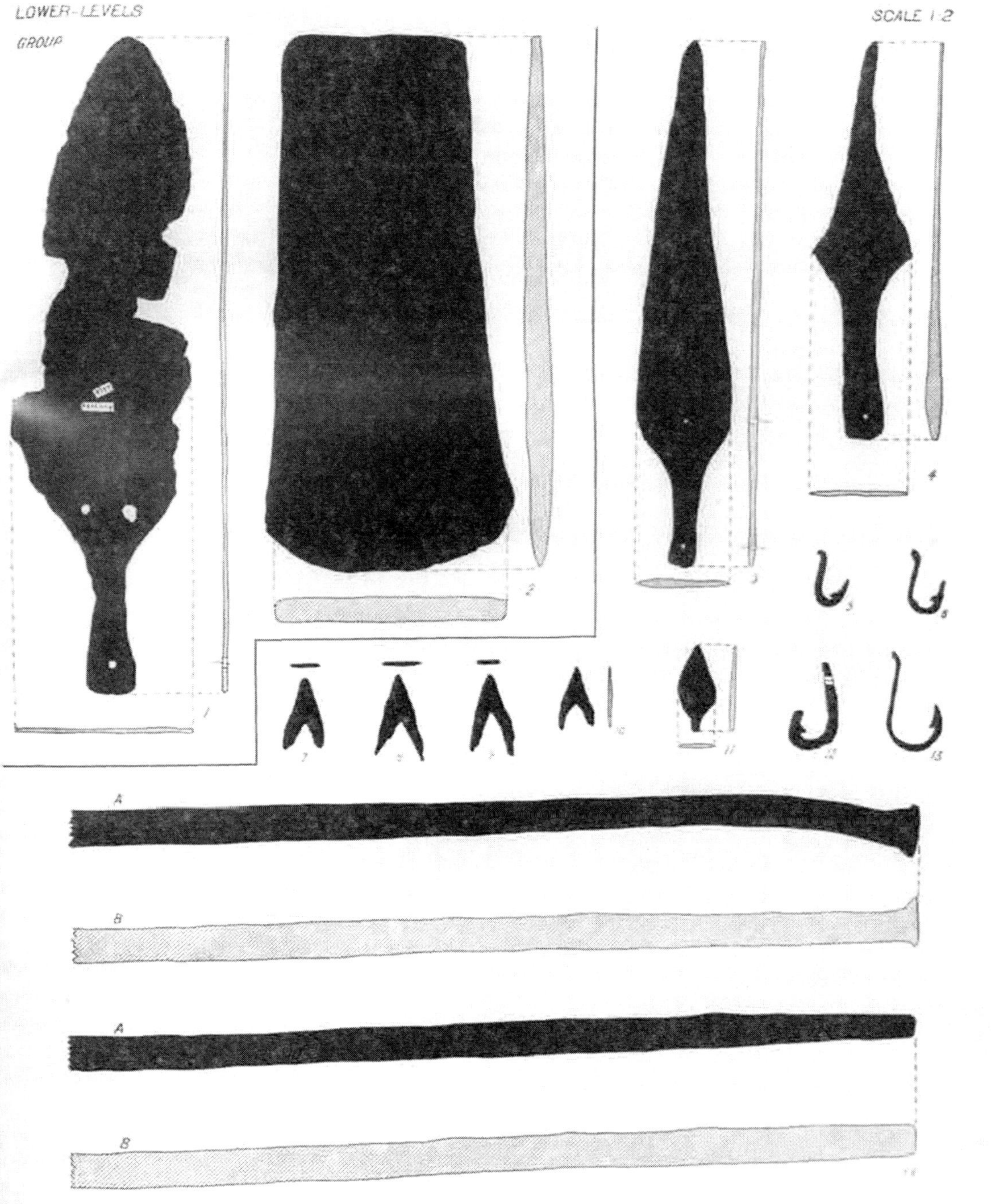

COPPER AND BRONZE TOOLS AND UTENSILS

SCALE 1:2

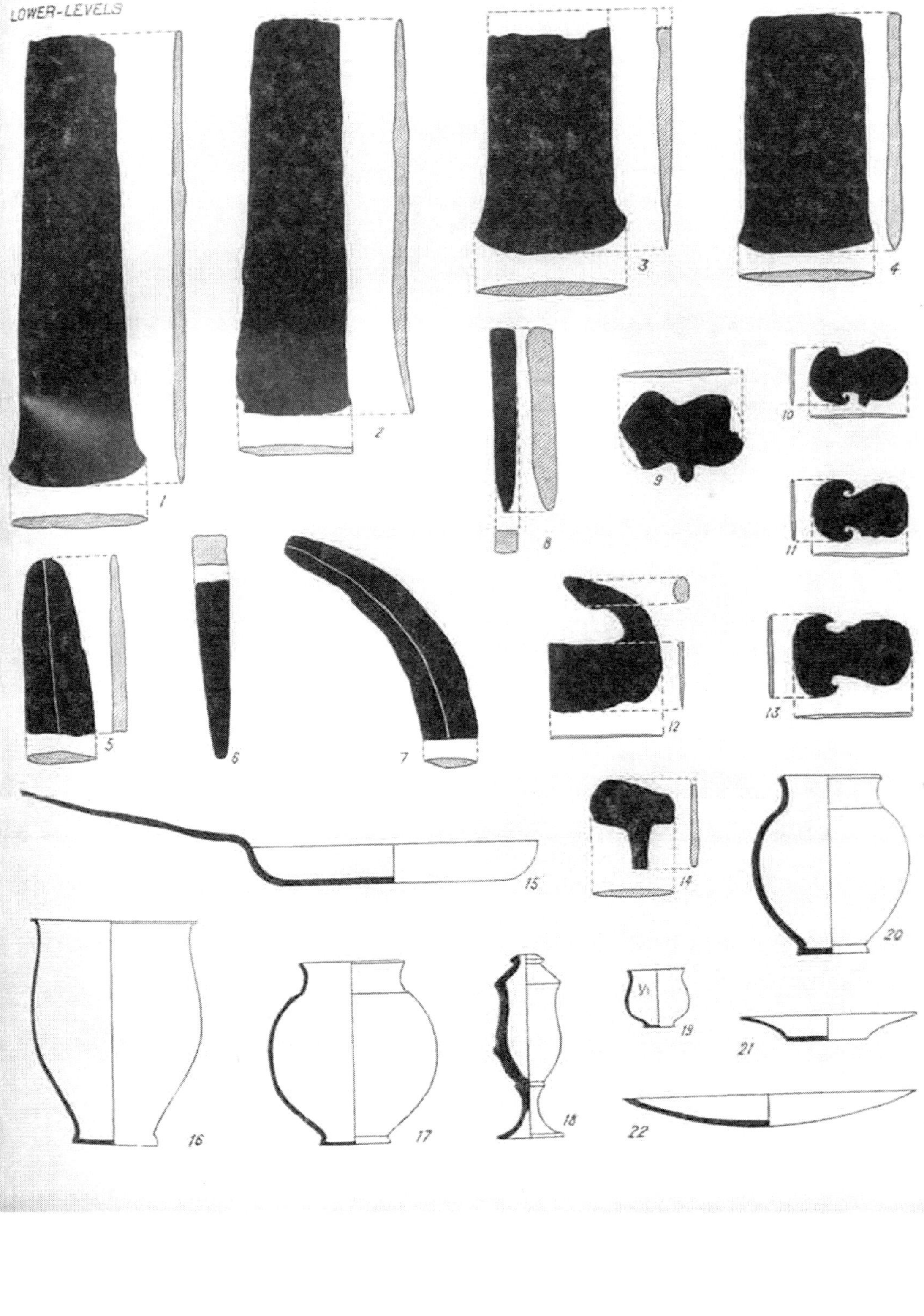

COPPER AND BRONZE KNIVES AND DAGGERS.

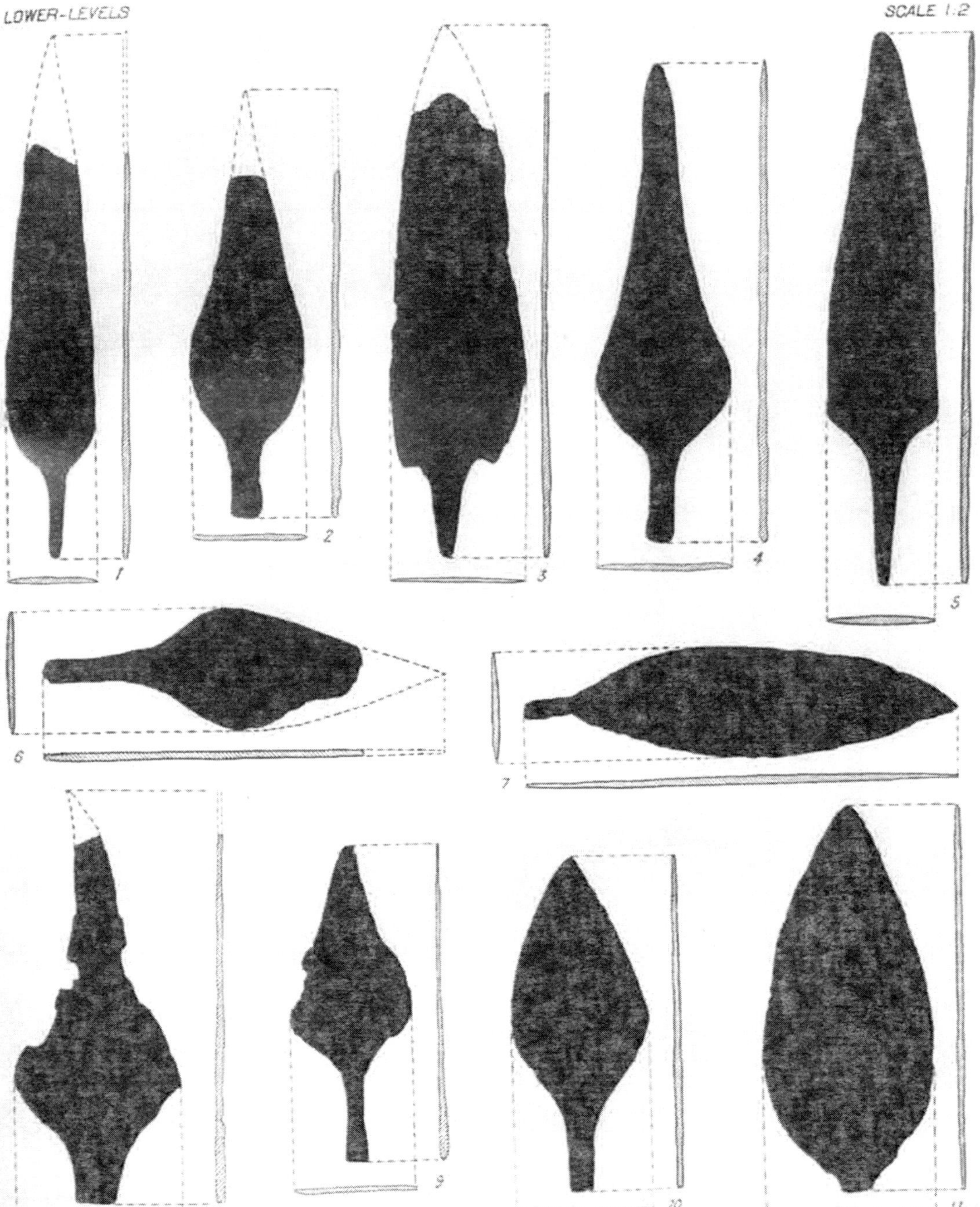

COPPER AND BRONZE TOOLS AND OTHER OBJECTS.

LOWER-LEVELS SCALE 1:2

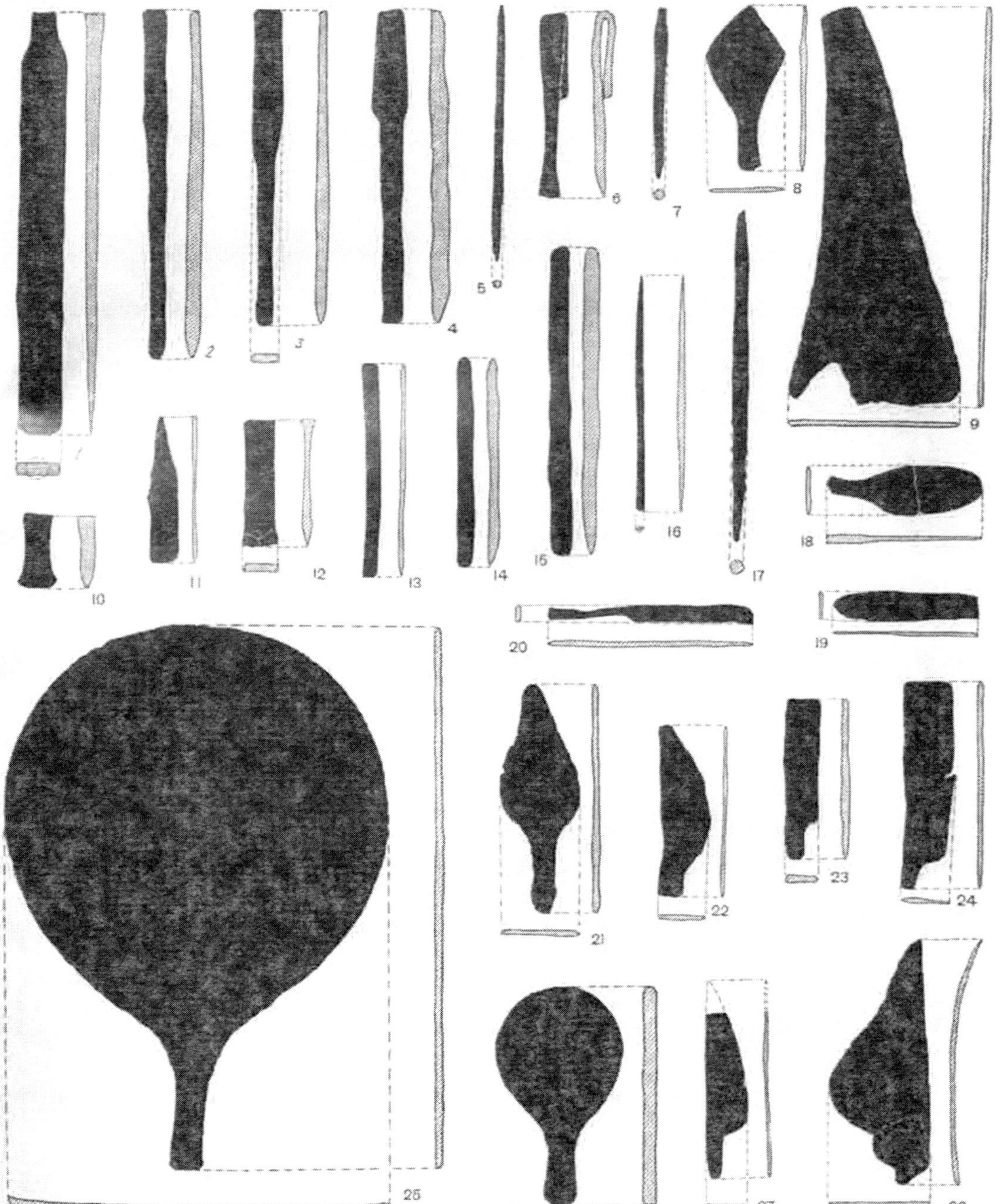

MOHENJO-DARO, DK. AREA, G. SECTION.
COPPER AND BRONZE TOOLS AND IMPLEMENTS. LOWER LEVELS.

MOHENJO-DARO, DK. AREA, G. SECTION.
COPPER AND BRONZE UTENSILS AND OTHER OBJECTS

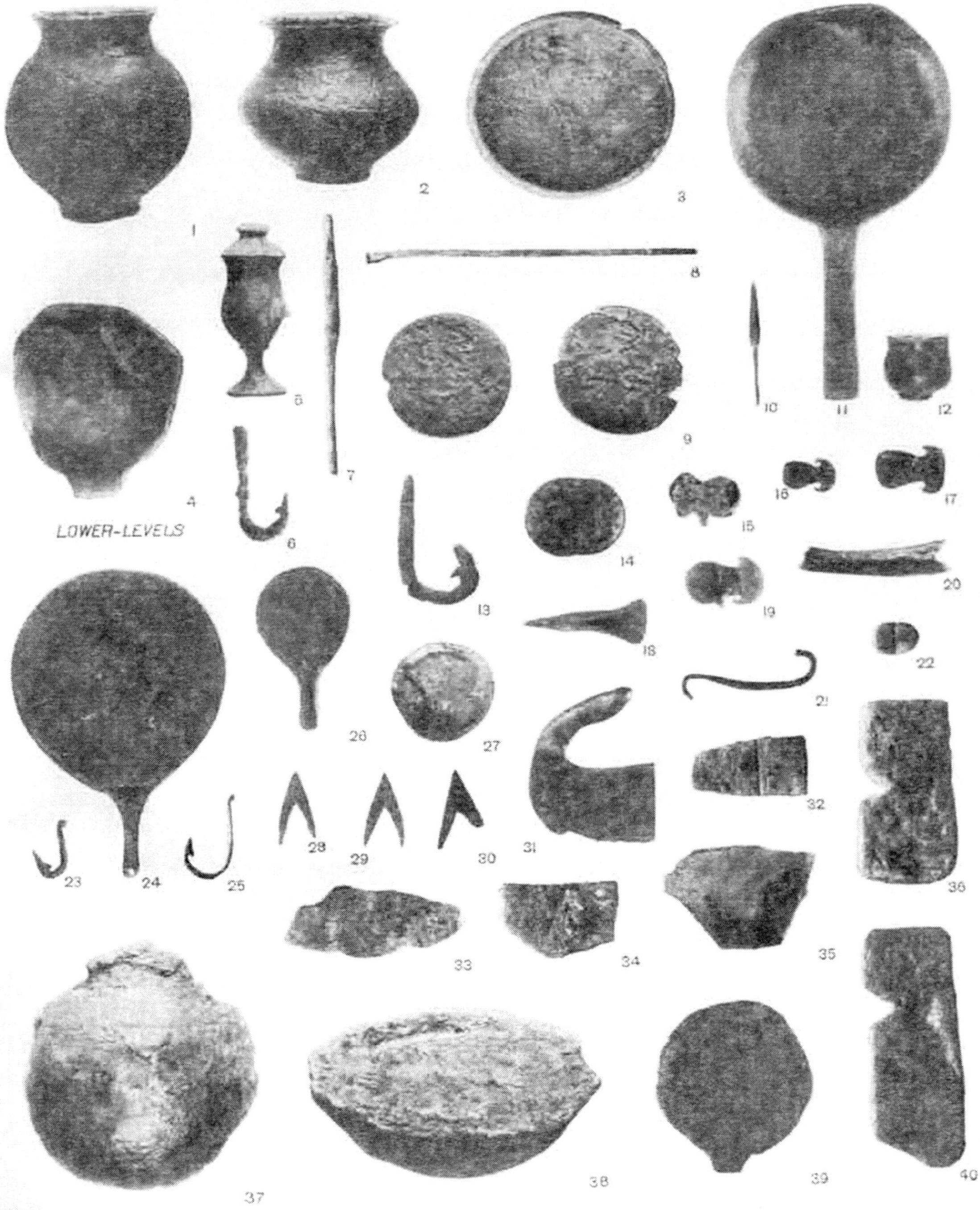

COPPER AND BRONZE TOOLS AND IMPLEMENTS.

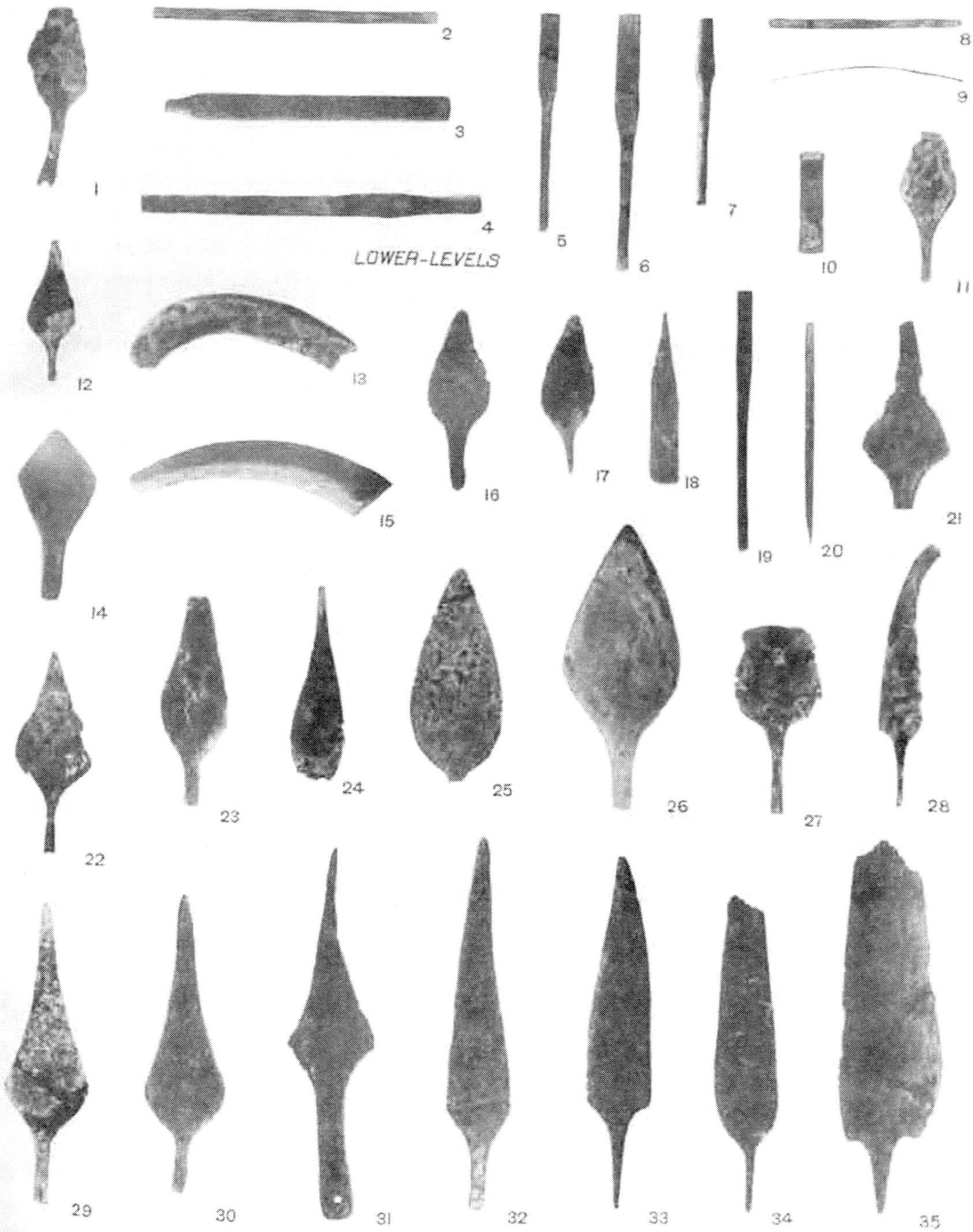

BEADS.

UPPER LEVELS

SCALE. 1:1.

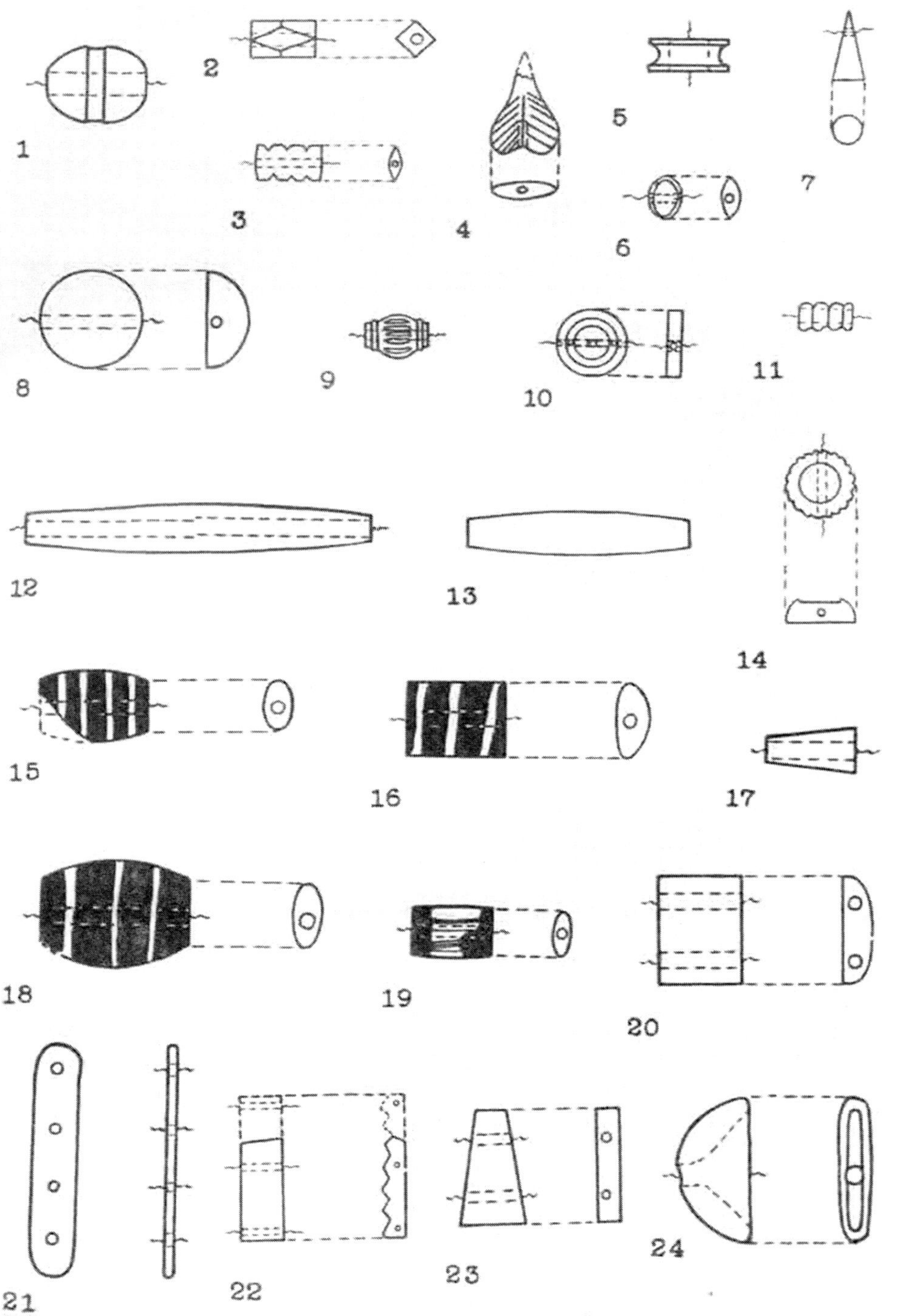

MOHENJO-DARO, DK. AREA. G. SECTION.
BEADS, CONES AND OTHER OBJECTS.

PLATE CXXXV.

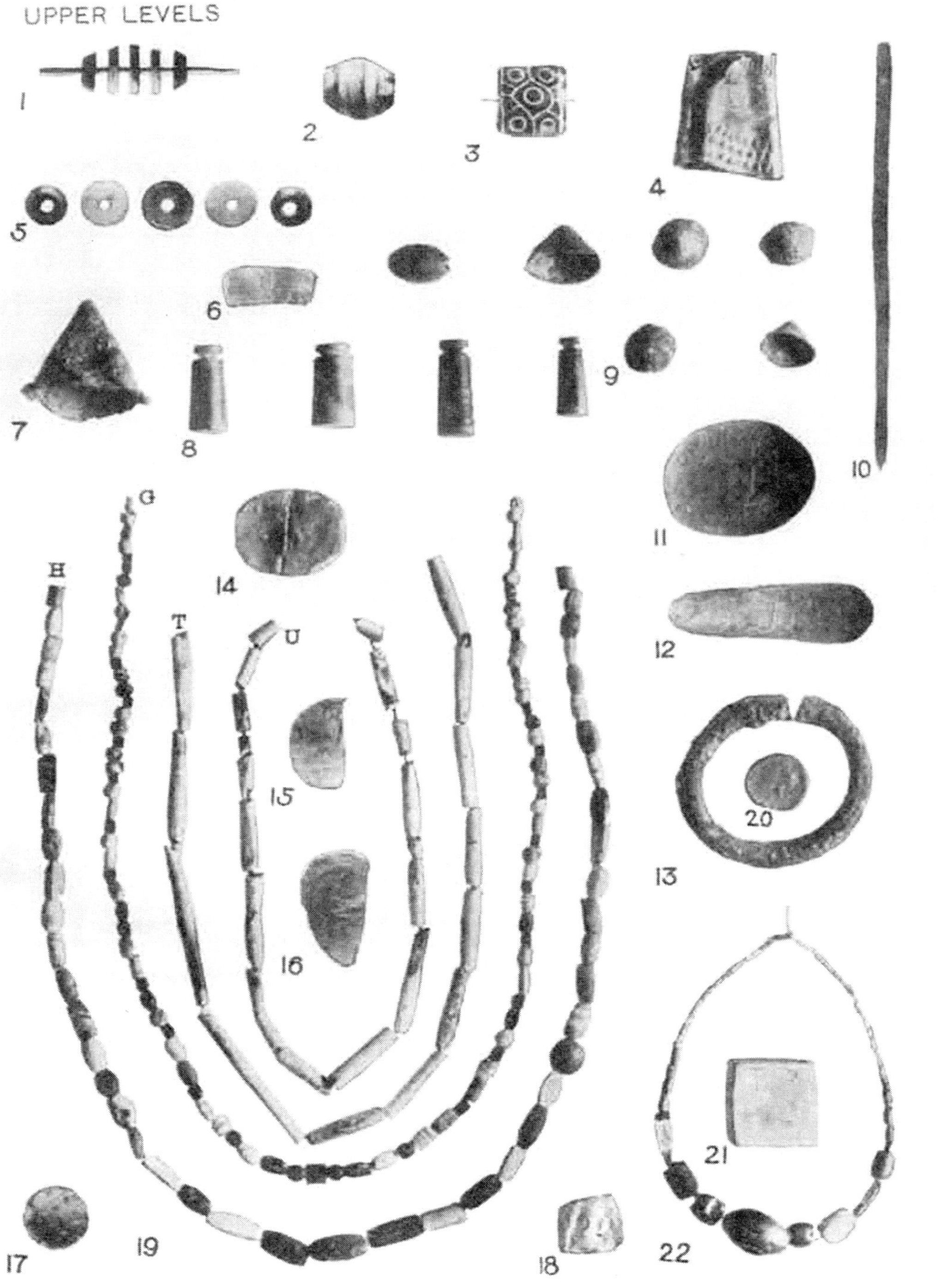

MOHENJO-DARO: DK. AREA, G. SECTION.
BEADS, AMULETS AND OTHER OBJECTS.

MOHENJO-DARO, DK. AREA, G. SECTION. PLATE CXXXVII

GAMESMEN AND BEADS.

UPPER-LEVELS SCALE 1:2

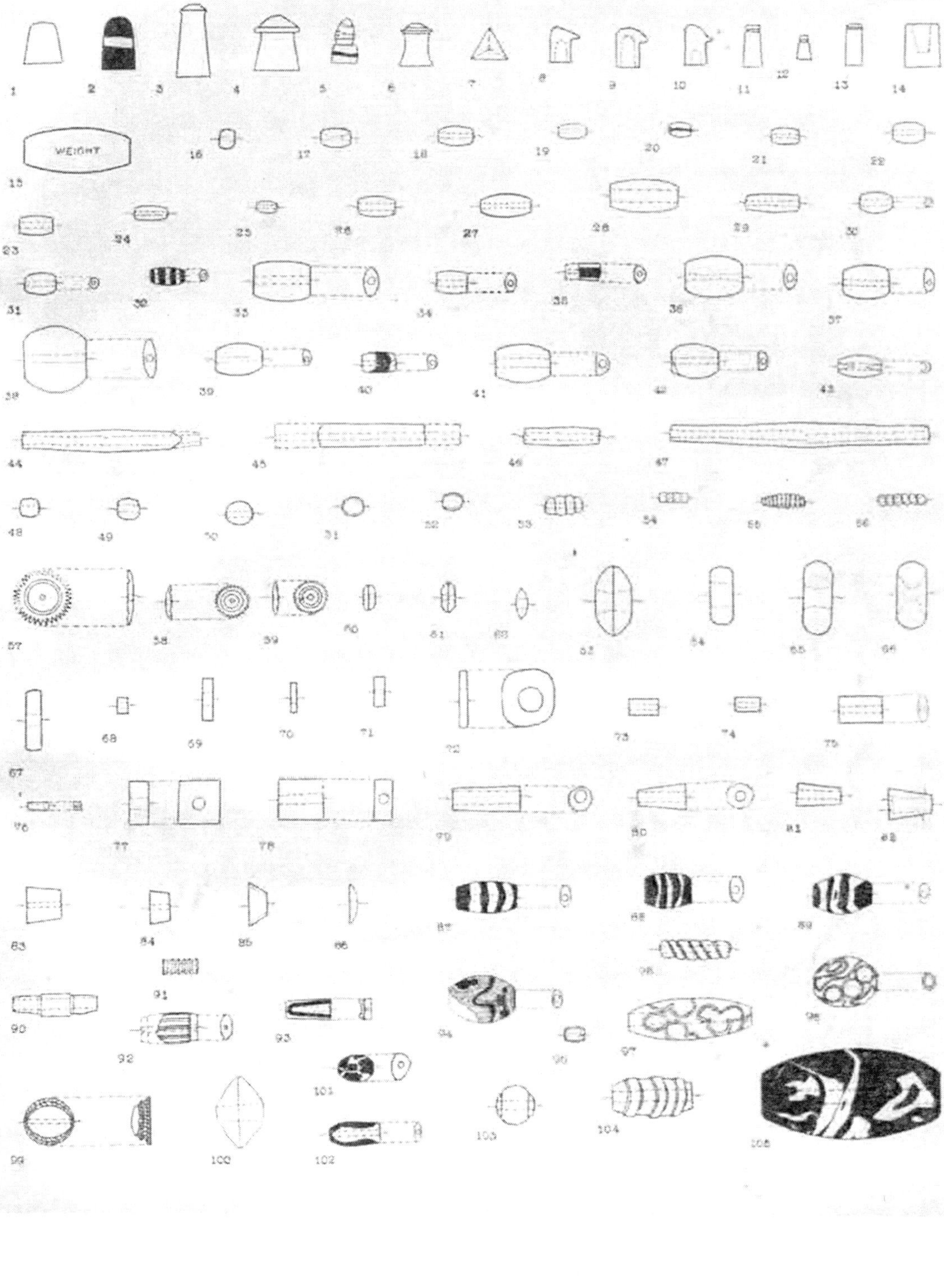

MOHENJO-DARO, DK. AREA, G. SECTION.

BEADS, IVORY DICE AND OTHER OBJECTS.

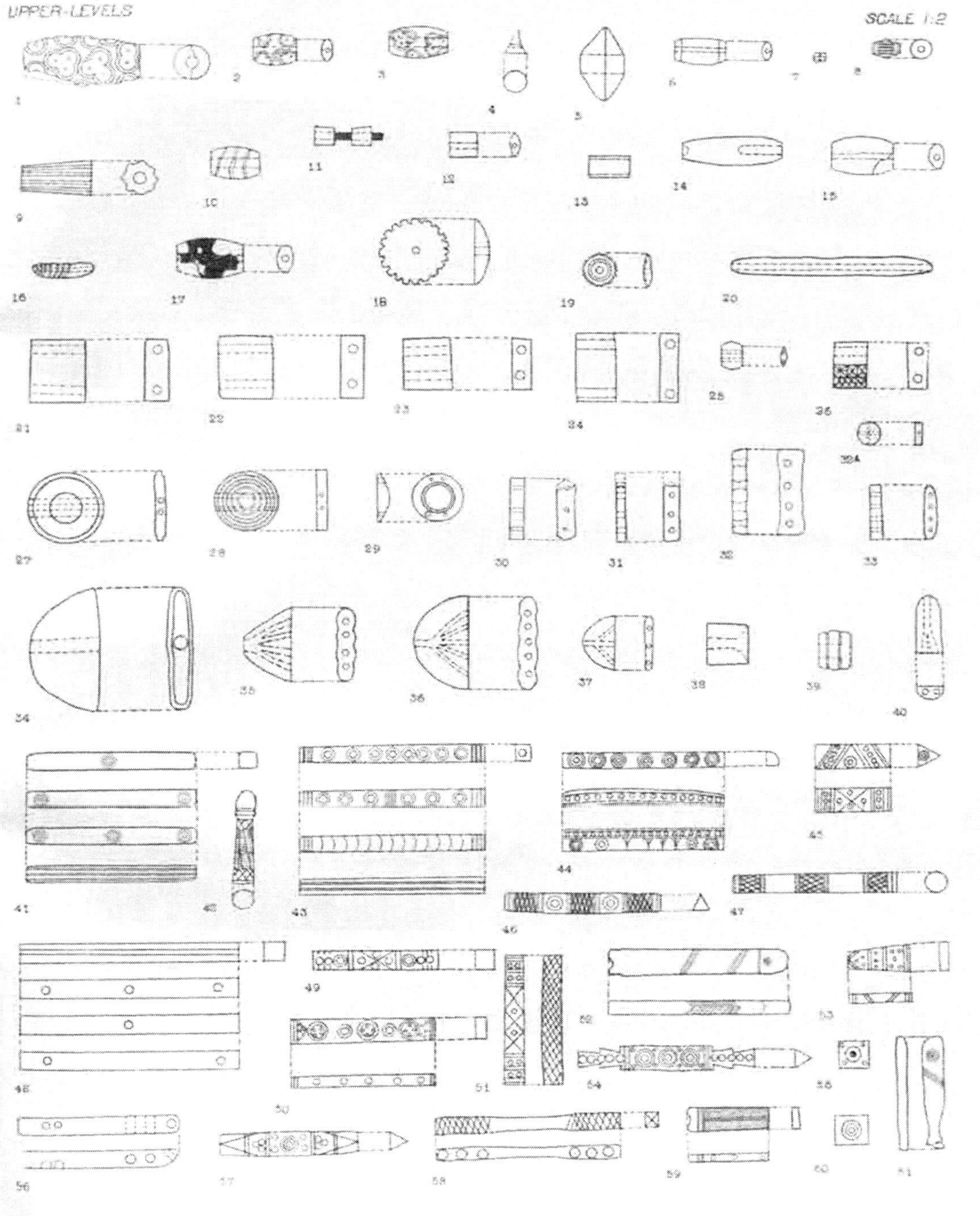

MOHENJO-DARO, DK. AREA, G. SECTION.

GAMESMEN, BANGLES, BEADS, ETC.

LOWER LEVELS

SCALE 1:2

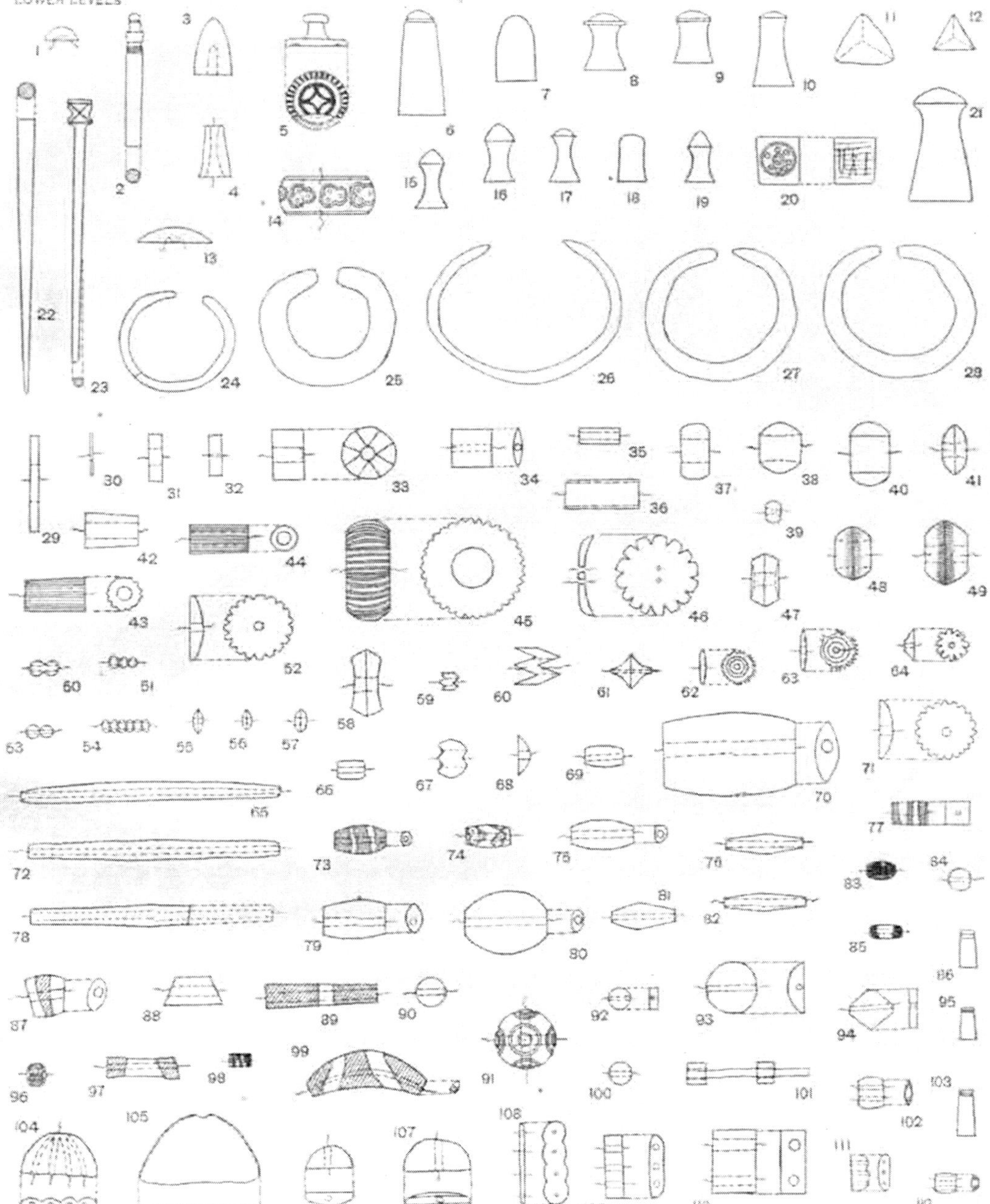

MOHENJO-DARO, DK. AREA, G. SECTION.
MISCELLANEOUS OBJECTS.

UPPER LEVELS

MOHENJO-DARO. DK. AREA, G. SECTION.
INLAYS, IVORY OBJECTS, SHELLS, POTTERY, BEADS, Etc.

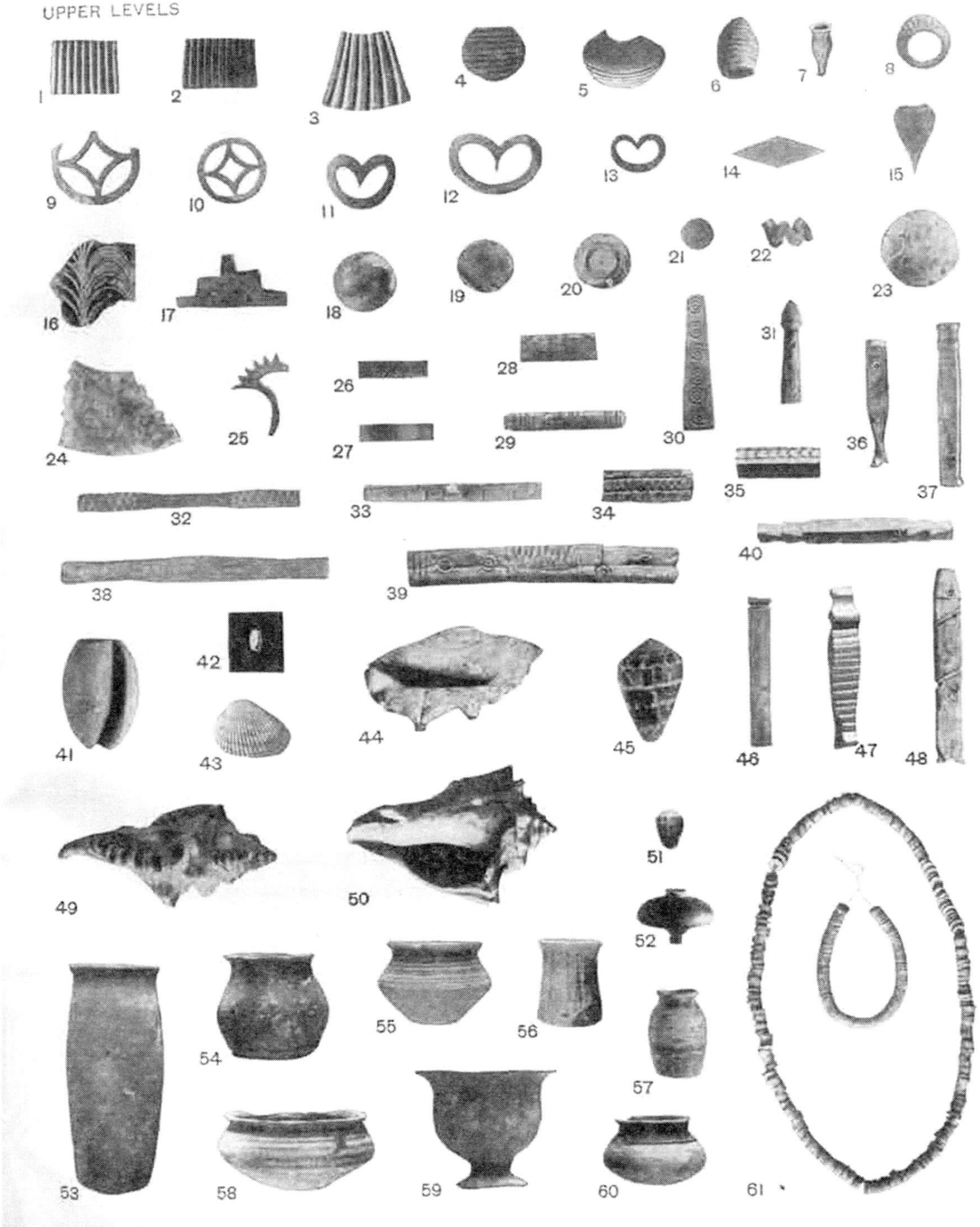

MOHENJO-DARO, DK AREA, G SECTION.

BULLAE, BUTTONS, BRACELETS, INLAY AND OTHER OBJECTS, LOWER LEVELS

STONE VESSELS, IVORY DICE AND OTHER OBJECTS.

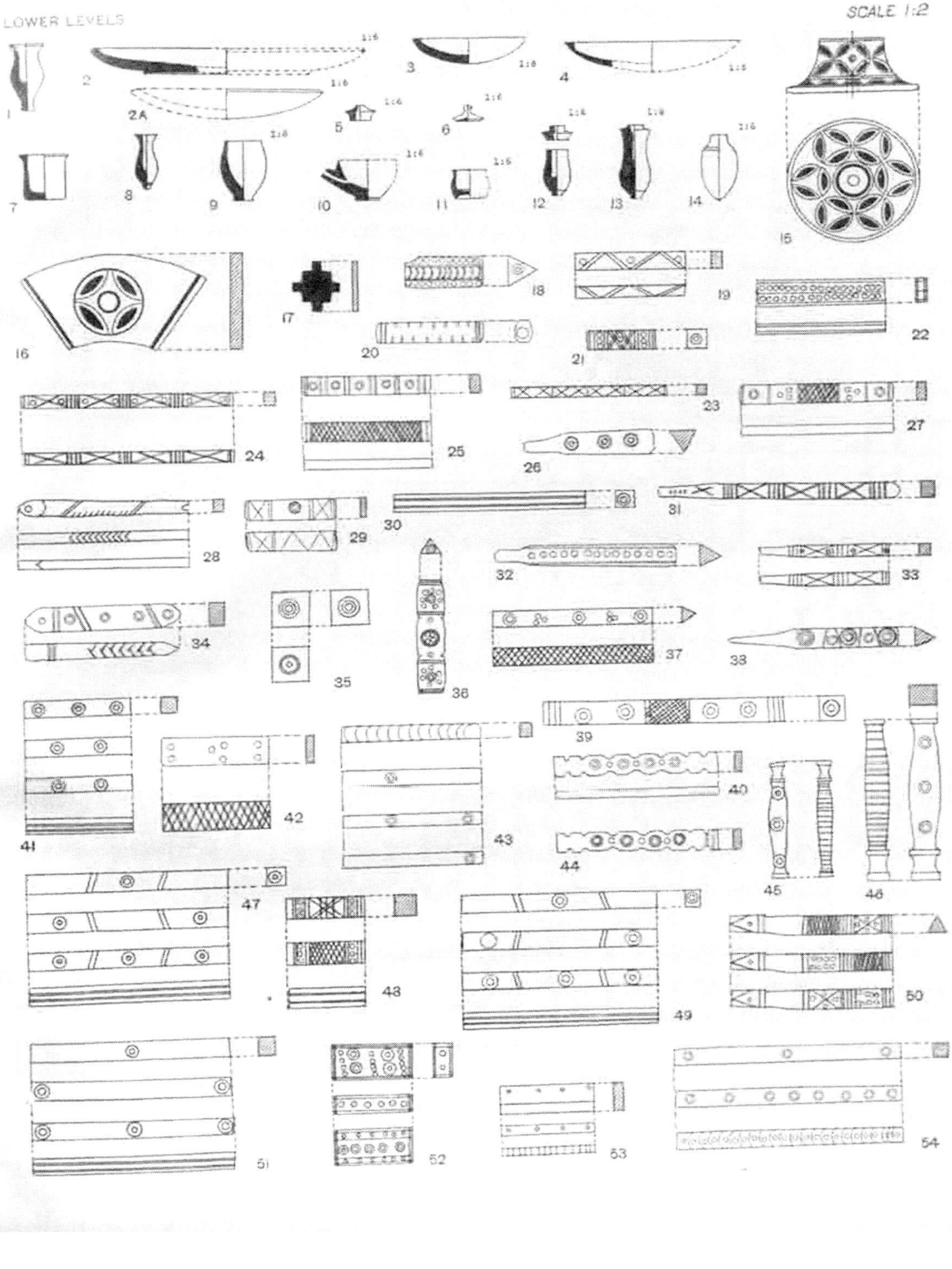

MOHENJO-DARO. HR. AREA.

RING-STONES.

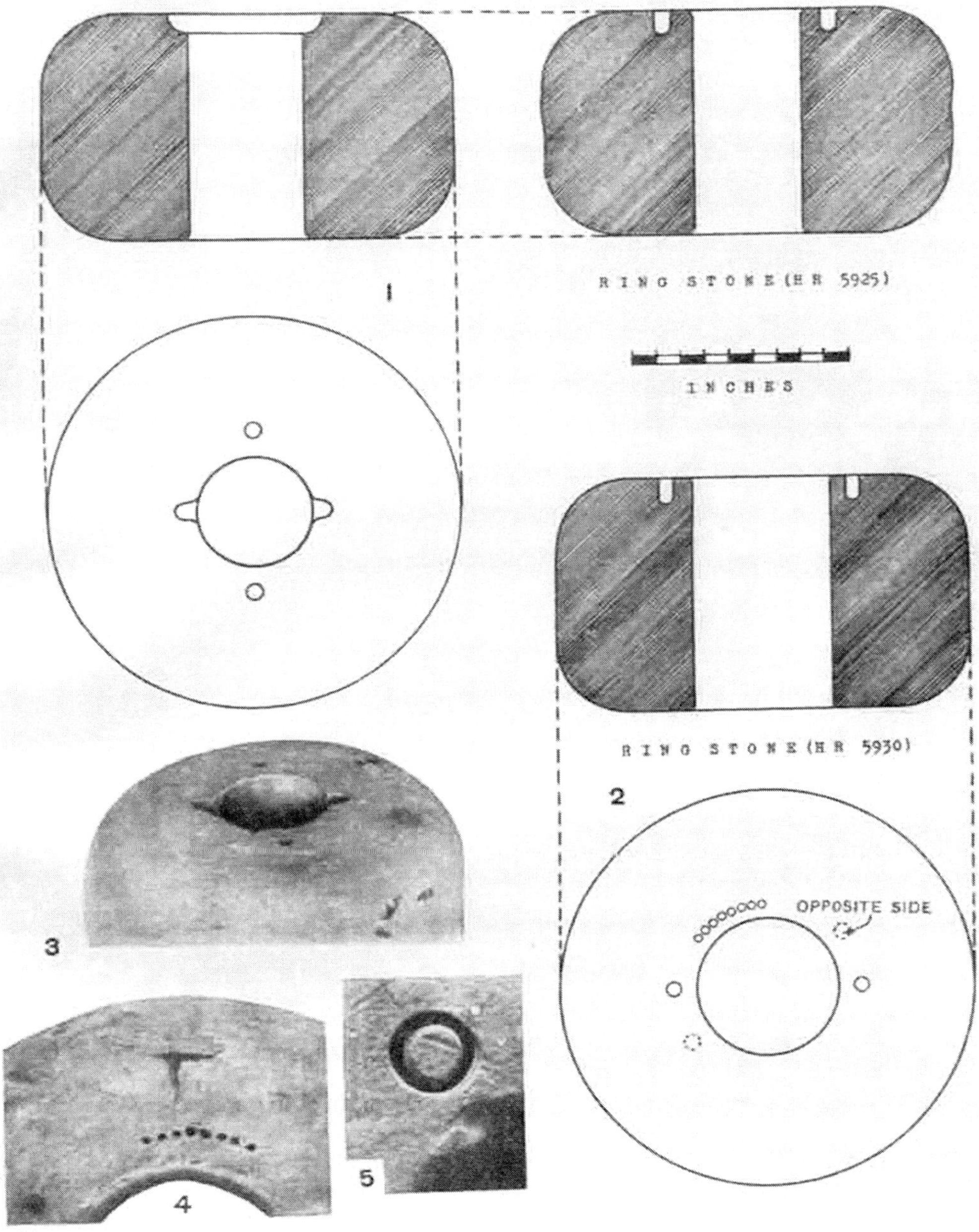

MOHENJO-DARO, DK AREA, G. SECTION. SKULLS.

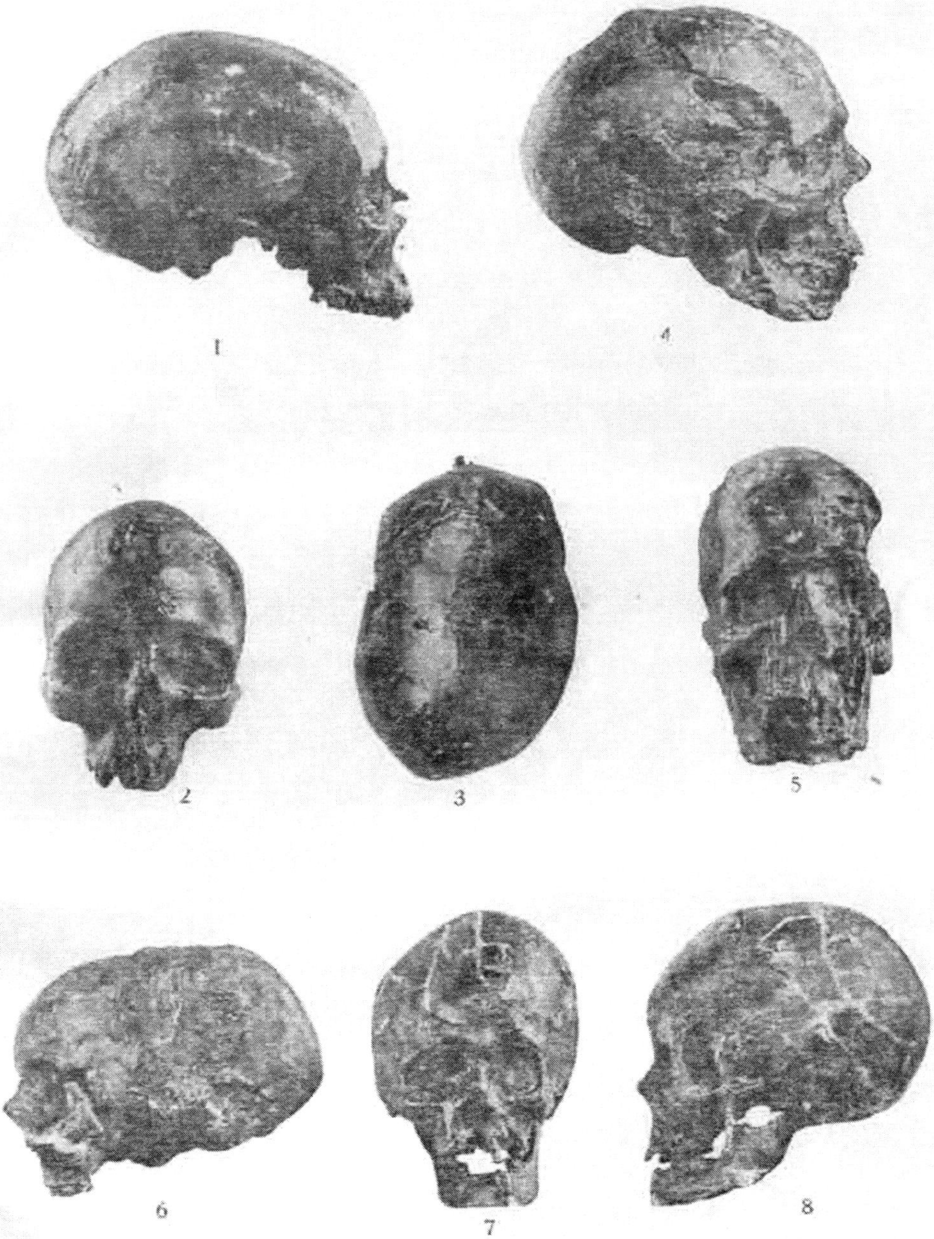

FIG. 1. NORMA LATERALIS OF M. 27.
FIG. 2. NORMA FRONTALIS OF M. 27.
FIG. 3. NORMA VERTICALIS OF M. 27.
FIG. 4. NORMA LATERALIS OF M. 28.
FIG. 5. NORMA FRONTALIS OF M. 28.
FIG. 6. NORMA LATERALIS OF M. 40.
FIG. 7. NORMA FRONTALIS OF M. 32.
FIG. 8. NORMA LATERALIS OF M. 32.

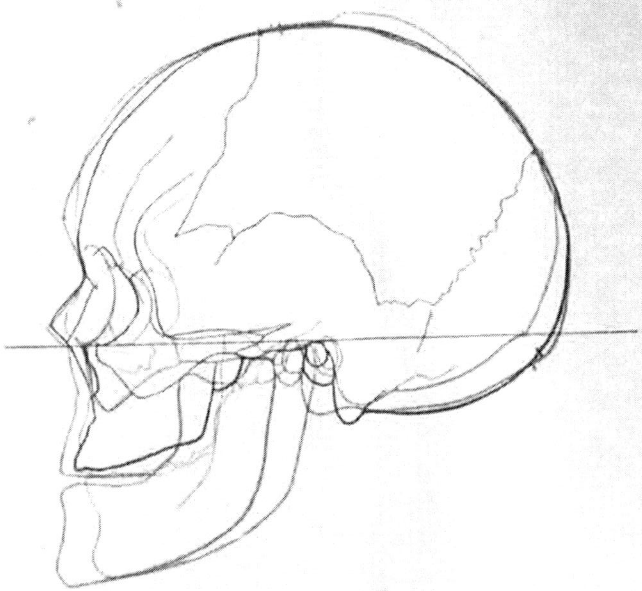

Fig. 1.

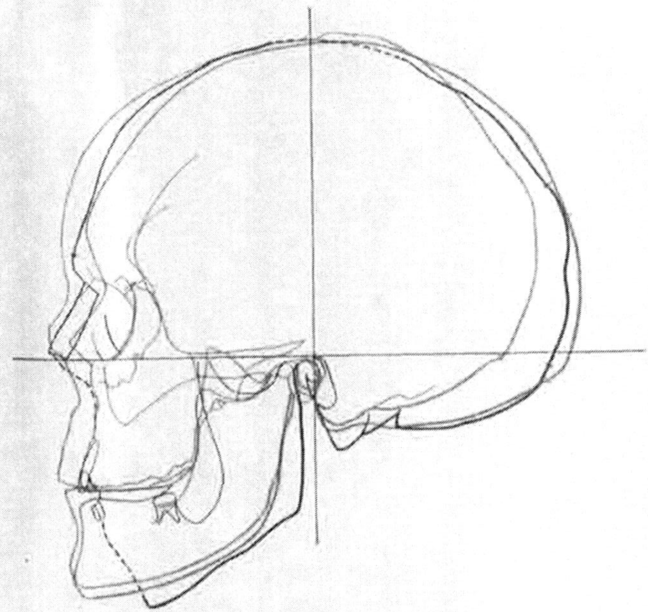

Fig. 2.

Mean of M. 27 and 28 ————
Mean of M. 2 and 11 ————

M. 28 ————
Mean of M. 2 and 11 ————

Published by

Facsimile Publisher
Ashok Vihar, Delhi-110052 (India)
Website: www.facsimilepublisher.com

Distributed by

Gyan Books Pvt. Ltd.
5, Ansari Road, Daryaganj, New Delhi-110002 (India)
E-mail: books@gyanbooks.com, Website: www.gyanbooks.com

2 222000 002435